Teaching 6–12 Math Intervention

W0081248

This practical resource offers a classroom-tested framework for secondary math teachers to support students who struggle. Teachers will explore an often-overlooked piece of the math achievement puzzle: the gatekeeping cycles of mathematics and the importance of teachers' own expectations of students. The immediately applicable strategies in this book, developed through the author's work as a math intervention teacher, intervention specialist, and instructional coach, will give teachers the tools to help students overcome math anxiety, retention struggles, and even apathy. Beginning with a deep dive into the gatekeeping cycles to help teachers better understand their students who struggle, the book then walks teachers through the five-part B.R.E.A.K. it™ Math Intervention Framework: Build Community, Routines to Boost Confidence, Engage Every Student, Advance Your Expectations, Know Students' Level of Understanding. Educational research, personal anecdotes from the author's own classroom, and examples from case study teachers are woven into each chapter, leading to clear action items, planning strategies, and best practices that are accessible enough to accommodate all grade levels and schedules. The framework and activities in this book enable teachers to help students overcome math anxiety, create a safe math environment for 6–12 students, and ultimately increase achievement with effective research-based suggestions for working with students who struggle.

Find additional resources at www.gatebreakerbook.com.

Juliana Tapper is the Founder of CollaboratEd, a math Professional Development consultancy for schools across the US and state departments of education. She has experience as a math classroom teacher, Professional Learning Community (PLC) facilitator, professional developer, and instructional coach throughout urban classrooms in multiple states, as well as district level leadership.

Also Available from Routledge Eye On Education
(www.routledge.com/eyeoneducation)

There is No One Way to Teach Math: Actionable Ideas for Grades 6–12
Henri Picciotto and Robin Pemantle

Coaching Math Workshop: A Must-Have Collection of Planning Forms,
Checklists, Reflection Sheets, Observation Logs, and More
Nicki Newton

Fluency Doesn't Just Happen in Multiplication and Division: Strategies
and Models for Teaching the Basic Facts
Nicki Newton, Ann Elise Record, Alison J. Mello

Guided Math Lessons in Fifth Grade: Getting Started
Nicki Newton

Day-by-Day Math Thinking Routines in Fifth Grade: 40 Weeks of Quick
Prompts and Activities
Nicki Newton

Guided Math in Action: Building Each Student's Mathematical
Proficiency with Small-Group Instruction, Second Edition
Nicki Newton

Mathematics Teaching on Target: A Guide to Teaching for Robust
Understanding of All Grade Levels
Alan Schoenfeld, Heather Fink, Alyssa Sayavedra, Anna Weltman, Sandra
Zuniga-Ruiz

Try It! Even More Math Problems for All
Jerry Kaplan

Messing Around with Math: Ready-to-Use Problems that Engage
Students in a Better Understanding of Key Math Concepts
David Costello

Introducing Nonroutine Math Problems to Secondary Learners: 60+
Engaging Examples and Strategies to Improve Problem-Solving Skills
Robert London

Exploring Math with Technology
Allison W. McCulloch and Jennifer N. Lovett

Teaching 6–12 Math Intervention

A Practical Framework To Engage Students Who Struggle

Juliana Tapper

Routledge
Taylor & Francis Group

NEW YORK AND LONDON

Designed cover image: © Getty Images

First published 2025
by Routledge
605 Third Avenue, New York, NY 10158

and by Routledge
4 Park Square, Milton Park, Abingdon, Oxon, OX14 4RN

Routledge is an imprint of the Taylor & Francis Group, an informa business

© 2025 Juliana Tapper

The right of Juliana Tapper to be identified as author of this work has been
asserted in accordance with sections 77 and 78 of the Copyright, Designs and
Patents Act 1988.

All rights reserved. No part of this book may be reprinted or reproduced or
utilised in any form or by any electronic, mechanical, or other means, now
known or hereafter invented, including photocopying and recording, or in any
information storage or retrieval system, without permission in writing from the
publishers.

Trademark notice: Product or corporate names may be trademarks or registered
trademarks, and are used only for identification and explanation without intent
to infringe.

ISBN: 9781032767017 (hbk)
ISBN: 9781032766980 (pbk)
ISBN: 9781003479703 (ebk)

DOI: 10.4324/9781003479703

Typeset in Palatino
by KnowledgeWorks Global Ltd.

Find additional resources at www.gatebreakerbook.com or scan the QR code!

Dedication

To my girls, A and O, may you find confidence in mathematics always.

To my husband, Stephen, you are more than I could have ever asked for or imagined. Thank you for your love and support.

To my mom and to my dad, thank you for always being excited for and supportive of my endeavors in your own unique ways.

To the teachers who have welcomed me or my strategies into their classrooms, thank you.

Lord, thank you for every twist and turn in my journey, every high and low in my classroom. Each one has led me to this moment and this book. I am forever humbled and grateful.

Dedication

Contents

Meet the Author

I'm a former high school math intervention and Algebra 1 teacher and co-teacher in South Los Angeles, East San Jose, and Denver. I truly understand what it feels like to teach a class full of students who have struggled with math for years and years. I also know how alone and solitary teaching the math intervention class can be without any concrete resources or guidance. However, I kept at it and tried and tweaked and retried and re-tweaked everything about my classroom until my intervention students were performing as well as – and in some cases better than – their "grade level" peers.

When I became a district math coach and TOSA (Teacher On Special Assignment) I led professional development for math and special education math teachers across thirteen urban high schools and began to notice themes about effective strategies for students who struggle. In 2018 I left the classroom to found CollaboratEd Consulting because I was frustrated with the lack of support, resources, and quality professional development for math teachers that work with students who struggle. I have been fortunate enough to provide effective and realistic professional development and coaching for schools, districts, and state departments of education across the country. I'm also an avid presenter at national conferences like NCSM (National Council of Supervisors of Mathematics), NCTM (National Council of Teachers of Mathematics), ASCD, and Learning Forward.

Introduction: Not The Mediocre Professional Development Book You're Used To

Room 223. That was my first classroom at Animo Watts Charter High School in South Los Angeles. I somehow snagged a corner classroom on the second floor as a first year teacher. Outside the line of windows along the back of my classroom was Magic Johnson Park with big green trees and a nice little pond that attracted a large number of Canadian geese. I was as ready as I could be. My teacher desk was decorated with my college and grad school flags, my bulletin boards were neatly outlined and ready to showcase student work, and the copies were made for my first day *get to know you* activities. I had spent the year prior student teaching at nearby Compton High School and I was excited to stay in South LA for my first full time teaching year. Every student at my school was a student of color – primarily Black and Brown – and receiving free or reduced lunch. I am a white cis woman who grew up incredibly privileged just twenty-two miles west of South LA in sunny Santa Monica. I had previously only passed through South LA on my way to Disneyland, but between my student teaching at nearby Compton High School and now my first year in neighboring Watts, I found myself spending more time in South LA than I ever imagined.

I don't remember much from my first day beyond struggling with students talking over me, giggles to ease my discomfort, and hanging out with my math twin, Sarah, in the teachers' lounge at lunch (the name was tongue-in-cheek: Sarah is Asian and 5'0" and I'm white and 5'10"; however, since we were the only 9th grade Math Support and Algebra 1 teachers we were lovingly referred to as the math twins). By day three, I wanted to quit. I had planned to do what I had experienced in my high school years: warm up,

DOI: 10.4324/9781003479703-1

homework review, lecture, worksheet, assign homework. However, many of my students had experienced math trauma, had math anxiety from repeatedly failing math, and were stuck in systemic gatekeeping cycles of mathematics (more about this in Chapter 1) and despite my good intentions, my classroom was chaos. Talking while I was lecturing, goofing off with friends during worksheet time, getting up to sharpen pencils and stopping for a chat with friends on the way back, shouting out answers, heads down on desks, students staring at the board with a confused haze, a constant stream of students asking to go to the bathroom, the crumpling of the worksheet, the refusal to take notes, throwing homework away on their way out the door. There were a few students who were doing their absolute best to stick with me, learn, and complete assignments, but unfortunately it was not the majority. It was exhausting, demoralizing, and the hardest year of my life. I felt like a complete and utter failure as a teacher.

By my second year, everything changed. Same school, different classroom (I got moved to the first floor this year), same park out my window, same geese, but this time I had found what worked for my students. Instead of chaos, there was productivity. Instead of off task talking, there was mutual respect and on task math discourse. Instead of talking to myself during lessons, there was 100% engagement and participation. Instead of zero work completion, there was a collaborative work ethic. Instead of dreading grading my quizzes, there was praise and joy when my students began earning some of the highest scores on our district benchmark exams. And although every day was still challenging, I absolutely loved coming to work every single day.

Inside this book are the strategies and mindset shifts that made my second year so different than my first. I'm sharing them l with you here because making sure every student succeeds and achieves at high levels of mathematics cannot wait.

Who Is This Book For?

This book is for any 6–12th grade math teacher with students who struggle with math in their classroom. General education teachers, special education teachers, interventionists, co-teachers, anyone who teaches students who have been historically unsuccessful in math, whether novice or veteran. In this book you'll meet brand new teachers who have wowed their administration with effective instruction and you'll meet twenty-year veteran teachers who were exhausted from the post pandemic apathy and who, through these methods, finally found enjoyment in teaching again.

This is a book about teaching math to students who struggle with math, to students who have been historically unsuccessful in mathematics, and who lack confidence and interest in math. If you teach honors math or don't have many students below grade level, this may not be the book for you. Every strategy and resource shared in this book is created with "those students" in mind; the students who come in disengaged from the start, the students who are reluctant to even get started, the students who goof off and act out as a way to cover up their struggles and trauma with mathematics.

This book is also for administrators. While this book is designed to be handed to any teacher to help support them in their work towards improving math outcomes for students, it works best when read as a department. If you're an administrator who has purchased or wants to purchase multiple copies of this book for your math department, you can get access to additional support materials for PLCs or department wide book studies at www. teams.gatebreakerbook.com.

Finally, this book made it into your hands one of two ways: you bought it for yourself and you're eager for help, or your administrator or supervisor bought it for you and requested you read it. If you're in the eager camp, I'm thrilled you're here! I know first-hand how transformative these strategies can be in a classroom and I'm excited for you to dig in and see a transformation of your own. If you're in the mandatory reading camp, I'm also thrilled you're here. My hope is that you and I might connect through the pages that follow and that I can earn your trust enough for you to try some of these strategies in your classroom.

What Makes This Book Different?

Instead of a book full of theories being tested by researchers, every single strategy in this book has been tested by me in my high school math intervention classroom because if we are being honest, teaching math to a classroom full of students who have struggled to find success in math has a unique set of challenges. Unless you've actually had to be in front of that audience day in and day out, you have no idea what it is like and you certainly don't have any right to be telling the teachers who have this daily experience what to do in their classrooms. To understand how challenging it is to teach math to students who have failed math for multiple years you have to actually do it, no amount of observation, guest teaching, or research is adequate.

I actually know how it feels to teach a class full of high school students who have failed math every year since 5th grade. I know how it feels to look out into your classroom and see a sea of students with hoodies up, earbuds in, and phones out. I know the pain of grading five periods of tests and feeling totally ineffectual as a math teacher. I know because I've actually been a high school math intervention teacher. I've taught ninth grade Algebra 1, high school repeater Algebra 1, co-taught Integrated Math 1, EOC Algebra 1 test prep for 11th and 12th graders, double dose of Algebra 1, and some more remedial classes at three different high schools in South Los Angeles, East San Jose, and Denver.

I actually know how it feels to teach a class full of high school students who have failed math every year since 5th grade. I know how it feels to look out into your classroom and see a sea of students with hoodies up, earbuds in, and phones out.

Instead of a book about how to teach a specific skill like multiplication, fractions, or linear functions this book is about an approach to teaching any math skill or content in a way specifically designed for students who have struggled with mathematics for many years. The framework laid out in this book will be your guide to increasing engagement, achievement, and motivation of your students who have been historically unsuccessful in mathematics.

The purpose of this book is two-fold. First, to provide you with the background knowledge to understand the *gatekeeping* nature of mathematics education. Second, to provide a practical path forward using math intervention strategies that have been proven effective in math classrooms around the U.S. In this book I will give you a step by step framework to become a *gatebreaker* so that your students who have historically struggled with math can thrive.

Why Me?

After a few more years in the classroom, a few years as a district math coach and teacher on special assignment (TOSA) in East San Jose, and then another year back in the classroom in Colorado, I decided I wanted to start sharing these methods that I had fine-tuned with other math teachers of students who struggle. Over the years I was forced to attend some really bad professional development. I sat in trainings that felt like a complete waste of time for me as a math teacher who really only taught students who had been historically unsuccessful in mathematics and I was tired of it. I felt the call to step out on my own and provide PD full time for my own company, CollaboratEd Consulting.

Since taking that leap of faith in 2018 I've provided PD for schools and districts across the U.S., I've established a partnership with the Colorado Department of Education and Colorado Department of Youth Services, and have been a presenter at national conferences like ASCD, Learning Forward, NCSM (National Council of Supervisors of Mathematics), and NCTM (National Council of Teachers of Mathematics) where I share my methods, strategies, and activities with other math teachers and leaders.

Over the years I have been fortunate enough to observe and coach in hundreds of secondary math classrooms and I've seen a variety of teaching styles, activities, management strategies, and student results. I've drawn some conclusions about what works to increase equitable outcomes when classrooms are made up of students who have been historically unsuccessful in mathematics. I began to realize that teachers who were achieving the most impressive student gains had five characteristics in common. These five characteristics make up the B.R.E.A.K. it™ Math Intervention Framework contained within this book: build community, routines to boost confidence, engage every student, advance your expectations, and know students' level of understanding. The framework has three distinct phases: student engagement, student achievement, and student motivation.

Let me assure you that it's not just me this framework works for, over the years I've collected anecdotal data of my own from hundreds of teachers who have implemented these strategies. At the end of every school year I give a survey to the teachers who have learned the framework and ask about their pass rates, engagement, and achievement before and after implementing the strategies they learned. On average teachers see 20% more students passing their class as well as a 46% increase in student engagement in their classrooms when they implement the strategies and methods from the framework within this book. What's really impressive is that many teachers achieve these gains within just one semester of implementation. These results are not just an average of one school or one district, they are the average of hundreds of 6–12th grade math teachers from all over the U.S. who have implemented the strategies in their classrooms. This framework has proven successful in rural schools, urban schools, suburban schools, in Title I schools, in facility schools, in juvenile justice schools, and more. This framework *is* for you too.

On average teachers see 20% more students passing their class as well as a 46% increase in student engagement in their classrooms when they implement the strategies and methods from the framework within this book.

All of it – my classroom experience, my district experience, my coaching experience – is in this book. Not only do I understand deeply what it takes to teach math intervention, I've discovered a path that works and I want to share it. Throughout this book you will hear my coaching voice, my teacher voice, and my learner voice as I take you on the journey to becoming a math gatebreaker. I wrote this book because I want to cut the learning curve for other math teachers with students who struggle and get right down to what works for our students who have been historically unsuccessful in mathematics. They don't have any more time to waste.

It Doesn't Have To Be This Way

Imagine a classroom where students who started the year "below grade level" walk in with confidence, participate in your warm up, engage in your lesson, actually practice math independently, and retain the information on tests and quizzes. Imagine how good it will feel to have some of the highest benchmark scores within your district. Imagine what it will feel like having an administrator walk out of your room during an observation and say, "Great job! Everyone was engaged!" Imagine if your planning time got more efficient and you could cut it in half. Imagine if your in-class formative assessment strategies were so dialed in that you didn't need to grade nightly homework. Imagine never feeling like you were talking to yourself all period long or that you're pulling teeth to get students to talk about math. That classroom is absolutely possible for you even if – make that, especially if – you teach students who have failed math for multiple years.

I know it might sound impossible, but the truth is, it is totally within your reach. The answers are within this book. Within the pages of this book you'll hear more about my personal classroom transformation, but more importantly you'll meet other teachers who have implemented the strategies: Sarah, Marcia, Melissa, Jessica, and Benny. They teach in title 1 schools and non-title 1 schools, they teach in urban schools, rural schools, and suburban schools, they teach middle school and they teach high school. And they have all seen the massive results you yearn for.

This work will not be easy nor will it be quick. You may spend an entire year just mastering the first two steps of the framework I present here, building community and implementing routines to boost confidence. While mastering these steps may not correlate to an immediate rise in test scores, if you've enabled students to have a positive experience in a math class after years of feeling like they don't belong here, you've made huge gains.

However, reading alone will not change anything, you must persevere and take action, after all that's what we're asking our students to do in our math classes, right? We all have unique classrooms, settings, and situations and if you find the strategies in this book are not working for you, connect with me on Instagram (@collaborated.with.juliana) so we can talk about it. The alternative to taking action with this book is to read a few pages here and there and let it collect dust on your desk or nightstand. Your students will continue to fail, be disruptive, and leave you feeling unfulfilled coming to a job you once loved. The choice is yours. Will you turn the page and begin the journey to the math classroom of your dreams?

1

Who Are Our Students Who Struggle?

I was a student who struggled with math. I didn't always like math. I wasn't always good at math. There was the time my parents hired a math tutor and he actually said, "You're so stupid. I can't believe you can't do this" while I froze during a multiplication worksheet he gave me. Then there was the time I walked into Algebra 1 class when I was in 8th grade. I wasn't prepared for it, but my mom knew taking Algebra 1 in 8th grade put me on the "college track" so she pulled the strings needed to get me into the class. I saw the warm up problem $2x + 6 = 10$ on the board and literally typed two times plus six into my calculator because I had no clue what a variable was.

There was a time when I thought my personal math experiences disqualified me from ever being a math teacher, but now I believe my experiences and struggles actually helped me be a better math teacher for my students. I can relate to them about the difficulty of learning math. I learned math even though it was hard for me and that's a message so many of our students who struggle need to hear.

Students Who Struggle: Defined

I've said that this book was written for 6–12th grade math teachers with students who struggle. I define students who struggle as any student who consistently finds it difficult to understand, engage with, or succeed in mathematics courses.

DOI: 10.4324/9781003479703-2

I define students who struggle as any student who consistently finds it difficult to understand, engage with, or succeed in mathematics courses.

Why Students Struggle

While there are many reasons why students struggle with math, there are seven characteristics that I've noticed most often in my classroom and within classrooms I visit.

- *Foundational Gaps:* When students struggle to learn a foundational skill (like multiplication, division, addition, or subtraction) that is needed for more advanced concepts.
- *Retention*: Students may seem to totally understand the lesson one day, but when they return the next day the skills didn't seem to stick.
- *Miscommunication:* A teacher explains a skill or concept in a way that doesn't resonate with or make sense to the student so they struggle to learn content from that teacher.
- *Systemic Inequities*: Students who are part of a specific demographic group that traditionally and historically has scored below other student demographic groups largely because of systemic inequities and institutional racism.
- *Negative Belief*s: Students may believe the myth of a "math gene" or hear adults in their life say, "it's okay, I wasn't good at math either" making math failure acceptable.
- *Math Anxiety:* Students may feel intense anxiety when they see numbers and struggle to problem solve, recall basic facts, or retain information.
- *Learning Disability*: Dyscalculia or other learning differences that make understanding and processing mathematical information more challenging.

How Students Struggle

When any of us struggles with something, it's nearly impossible not to feel stress or like we've experienced a bit of a trauma. When students struggle with math, the result is no different. And what do our bodies do when we're stressed or have experienced a traumatic event? Our primal brain takes over and we go into fight or flight. However, trauma research suggests there are actually four possible responses: fight, flight, freeze, or fawn (Taylor, 2022).

Here are some ways I've seen fight, flight, freeze, or fawn manifest in my math intervention classroom:

- *Fight*: Off task, disruptive, or disrespectful behavior instead of attempting the math. Students who are disengaged in math class from the start and put little to no effort into even getting started.
- *Flight*: Ditch class in order to avoid the risk of failure or go to the bathroom every single day to escape.
- *Freeze*: Quietly withdrawing from class because their body can't fight or flee from the situation, but all they can focus on is an intense sense of dread.
- *Fawn*: A student may put on a front to be participating in order to please you, but they are actually completely lost.

Why We Need to Intervene When Students Struggle With Math

Access and success with grade level content in mathematics is more important than any other subject. In 1997 the U.S. Secretary of Education released a white paper titled, *Mathematics Equals Opportunity*, which states:

> In the United States today, mastering mathematics has become more important than ever. Students with a strong grasp of mathematics have an advantage in academics and in the job market. The 8th grade is a critical point in mathematics education. Achievement at that stage clears the way for students to take rigorous high school mathematics and science courses—keys to college entrance and success in the labor force. (1) Students who take rigorous mathematics and science courses are much more likely to go to college than those who do not. (2) Algebra is the "gateway" to advanced mathematics and science in high school, yet most students do not take it in middle school. (3) Taking rigorous mathematics and science courses in high school appears to be especially important for low-income students. (4) Despite the importance of low-income students taking rigorous mathematics and science courses, these students are less likely to take them.
>
> (U.S. Department of Education, 1997, pp. 5–6)

This white paper makes it clear: Algebra is a gate that students must pass through to enter higher level math classes, and taking higher level math classes increases students chances of going to college as well as giving advantages in the job market. Algebra 1 teachers are not the only teachers responsible for opening the gate. Every teacher a student has leading up to Algebra 1 plays a role in what happens when that student reaches Algebra 1.

It's up to us to help all of our students have positive experiences and success in mathematics so that when they get to Algebra 1, the gate will open. This white paper also makes it clear that success in mathematics for low-income students is particularly important and draws attention to the fact that low-income students are less likely to enroll in higher level math courses. We will see shortly that being stuck behind a closed gate in mathematics kicks off a vicious cycle for many students, one that can lead to being caught in a pattern of low achievement for some and dropping out of school altogether for others. As math teachers, we must be aware of this gate and do everything in our power to open the gate and intervene on behalf of all students, but particularly students who are struggling to find success in mathematics.

The Gatekeeping Cycles of Mathematics

Throughout this book you will frequently see me use the term gatekeeper. When I first became a math teacher, I didn't know "mathematics gatekeeping" was a real thing, but now I know it is and I'm committed to breaking down as many gates as possible. We are going to explore examples of gates that connect to your classroom in more depth in this section, but as a quick introduction, Algebra 1 is likely the most notorious gatekeeper because the course acts as a gate to higher academic achievement and success. When students pass Algebra 1 they essentially unlock the gate to higher level math courses,

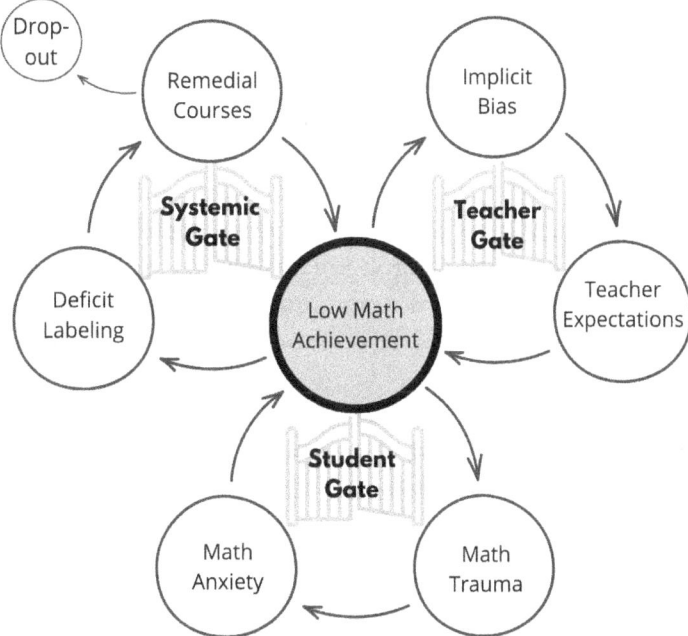

Figure 1.1 The Gatekeeping Cycles Of Mathematics: Made Up Of A Student Gate, Teacher Gate, Systemic Gate

those courses are needed for high school graduation and college acceptance. When students fail Algebra 1 they are essentially met with a closed gate, their options are only to keep repeating Algebra 1 until they pass it, wasting valuable time in high school and increasing the likelihood they drop out.

As shown in Figure 1.1 there are three cycles that make up this gatekeeping of mathematics and we can think of these cycles as gates. These gates are in the way of increasing outcomes for students who struggle with math in our current education system. There is a student gate, a teacher gate, and a systemic gate. Some students are stuck behind one gate, repeating the cycle over and over again and some students start within one cycle and get flung around only to be met with another gate as they find themselves stuck in yet another cycle. In the pages of this book I will teach you how to break these gates and help students who struggle to find success in mathematics and thrive. Before we can talk about addressing issues at the teacher and student level, we need to look at the system level, which means looking at the concept of equity.

In the pages of this book I will teach you how to break these gates and help students who struggle to find success in mathematics and thrive.

Math Equity

When we talk about low student math achievement and students who struggle with math we also need to acknowledge, accept, and understand that inequities in math education exist in this country. The National Center for Education Statistics releases a "Condition of Education" report every year. While the report shows a downward trend for math scores since 2012, it also reveals a disparity between demographics such as ethnicity, economic means, disability status, and language learning status. For context, in 2022, NAEP (National Assessment of Educational Progress) defined a score of 299 to be proficient and score of 262 to be a basic achievement level score (U.S. Department of Education, various years). The 2022 average math scores for 8th graders as measured by the NAEP tests are shown in Figure 1.2. There are four demographics we'll be taking a closer look at: ethnicity, income level, disability status, and English learner status.:

- ◆ *Ethnicity*: Here are the results for average 8th-grade NAEP scores: Asian students 306, White students 285, Hispanic students 261, American Indian/Alaska native 258, Black students 253. That is a fifty-three-point difference between Asian and Black 8th grade students' math scores (National Center for Education Statistics, 2024).

◆ *Poverty*: The same report shares scores for students from "high poverty schools" and "low poverty schools." In 2022, the average NAEP mathematics score for 8th-grade students in high-poverty schools was 271 and in low-poverty schools it was 293, a twenty-two-point difference (National Center for Education Statistics, 2024).

◆ *Disability Status*: A demographic sometimes left out of the equity conversation is students with learning differences. In 2022, the average NAEP mathematics score for 8th-grade students identified as a student with a disability was 243 while students not identified as having learning differences was 279, a thirty-six-point difference (National Center for Education Statistics, 2024).

◆ *English Language Learners:* Lastly, there is a disparity between multilingual learners and English only students with average scores landing at 241 and 277 respectively, another thirty-six-point difference (National Center for Education Statistics, 2024).

One bright spot in this challenging data is that the math gender gap seems to be narrowing with 8th-grade males having an average score of 275 and females having an average score of 273. However, the upsetting truth remains: Students of color, students from historically underserved backgrounds, students with disabilities, and multilingual students are scoring disproportionately lower – and

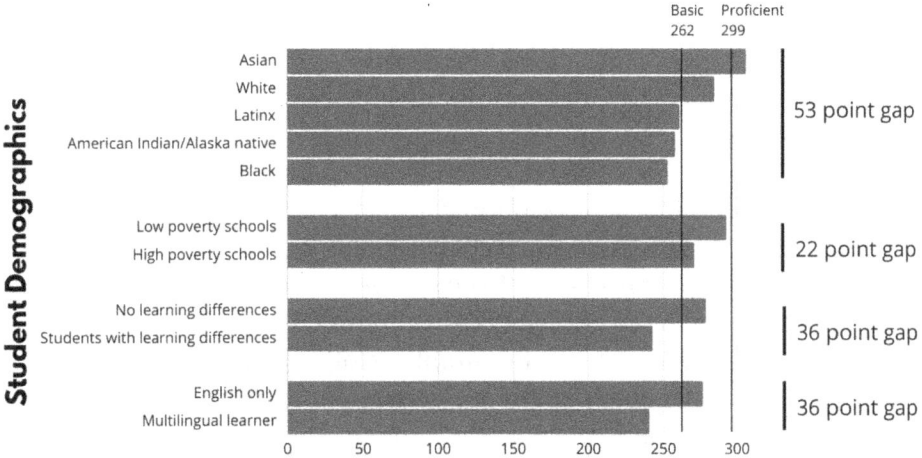

Avg NAEP Math Scores for 8th Graders in 2022

Figure 1.2 Bar Graph Of Average NAEP Math Scores For 8th Graders In 2022 By Demographics With Score Differences Identified

Created by author from data from National Center for Education Statistics (2023).

therefore struggling more – in math than white students, Asian students, higher-income peers, students without disabilities, and English only peers. Any book that discusses strategies for students who struggle must also make sure to look through the suggestions with an equity lens.

Let me be very clear that I am not saying every Black student will struggle with math, or every low-income student will struggle with math, or every student with an IEP will struggle with math, or every multilingual learner you have in your classroom will struggle with math. Demographics are not a students' mathematical destiny. What I am saying is that math inequities do exist in this country and we, as math teachers, must be aware of the role they play in why students struggle.

Demographics are not a students' mathematical destiny.

Math Equity Defined

In his book, *Equity Now*, Dr. Tyrone Howard defines equity as, "providing the necessary supports, interventions, and actions to correct past wrongs with the goal of justice and belonging for all students" (Howard, 2024, p. 20). The National Council of Teachers of Mathematics (NCTM) released a statement titled, "Access and Equity in Mathematics Education". They define math equity as:

> Creating, supporting, and sustaining a culture of access and equity require being responsive to students' backgrounds, experiences, cultural perspectives, traditions, and knowledge when designing and implementing a mathematics program and assessing its effectiveness. Acknowledging and addressing factors that contribute to differential outcomes among groups of students are critical to ensuring that all students routinely have opportunities to experience high-quality mathematics instruction, learn challenging mathematics content, and receive the support necessary to be successful. Addressing equity and access includes both ensuring that all students attain mathematics proficiency and increasing the numbers of students from all racial, ethnic, linguistic, gender, and socioeconomic groups who attain the highest levels of mathematics achievement.
>
> (NCTM, 2014, p. 1)

Both the NCTM position and Dr. Howard's definition of equity point to the systemic inequities present in our education system. They also both point to this notion of belonging. Equity means that all students, regardless of identified demographics, feel they belong in the highest math courses.

Here is what I believe we can take from these two definitions of equity to define math equity for this book: Equity in mathematics looks like not having ethnic, linguistic, gender, ability, and socioeconomic disparities in achievement data or course enrollment data. In order to achieve mathematics equity math teachers must understand, acknowledge, and address that inequities exist and learn strategies, supports, and interventions that will ensure all students are given what they need to belong in high level math courses and attain the highest level of mathematics achievement. This book attempts to give you tools, strategies, and ideas for supports and interventions that will lead to more equitable achievement outcomes.

Equity in mathematics looks like not having ethnic, linguistic, gender, ability, and socioeconomic disparities in achievement data or course enrollment data. In order to achieve mathematics equity math teachers must understand, acknowledge, and address that inequities exist and learn strategies, supports, and interventions that will ensure all students are given what they need to belong in high level math courses and attain the highest level of mathematics achievement.

Now that we have a deeper understanding that inequities do in fact exist in mathematics education, let's dig deeper into each of the gates that we must break in order to improve outcomes for all students who struggle with math.

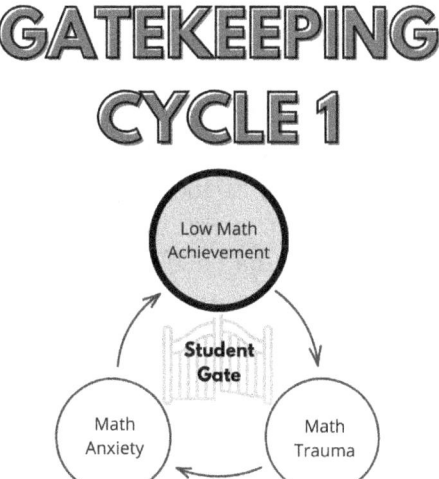

Figure 1.3 Gatekeeping Cycle #1: Student Gates – Math Trauma And Math Anxiety

Gatekeeping Cycle 1: Student Gate

The first gate in the way of improving outcomes for students who struggle is the student gate made up of math trauma and math anxiety.

Math Trauma

What exactly is math trauma? On her Math Therapy podcast, host Vanessa Vakharia defines math trauma as, "the lasting emotional impact that living through a distressing event relating to math, relating to math class, relating to your relationship with math, might have had" (Vakharia, 2023). There are many events that could be math trauma triggers for our students. Some of the most common math trauma events that I've seen in my students and the students of the teachers I support are:

- Getting an answer wrong in front of the class and having peers laugh
- Getting an answer wrong and being ignored or dismissed by the teacher
- Being called stupid or dumb in class (by a teacher or peers)
- Not finishing timed fact tests in the time provided
- Getting a test back covered in red ink
- Repeatedly failing math class

It might be a little uncomfortable realizing that our classrooms have created trauma for our students. While I believe all teachers are well meaning and no one intentionally sets out to traumatize students, the reality is that it has happened and now we have to do something about it if we want to help students get free from this gatekeeping cycle and begin to thrive in mathematics.

Many more students have experienced deeper trauma than we realize and while our job as teachers is not to process trauma with our students – instead we should connect them with trained counselors who can help them process trauma – it is important that we understand how trauma impacts students' mindset and behavior in our classrooms since math trauma is a real thing. While we tend to think of trauma as the "Big T Trauma" like neglect and abuse or singular traumatic events, the brain is impacted the same way when students experience a "little t trauma" like getting an answer wrong in front of the class, repeatedly failing the same subject, or countless other ways traumatic math event may have been experienced by a student. With trauma we know that "when students have experienced trauma, the brain is impacted, particularly the areas that support learning and development. If teachers are not aware and versed in this, the likelihood of students successfully learning as best as possible can be lessened" (Ferlazzo, 2022, paragraph 2). As math teachers we must understand that our class may be a trauma trigger for many of our students and if we want students to successfully learn mathematics as best as possible, we have to help them explore, express, and overcome any math trauma. In Chapter 3 I'll be sharing practical strategies to do this.

Throughout this book you'll see "Math Trauma Connection Points" are there to help you remember the role math trauma plays in our students' math experiences and how we can help them overcome it so that they can thrive.

Math Anxiety

Some teachers are relieved when I tell them math anxiety is a real thing. It's like they've known there has to be more than just students not liking their class, they knew it was something deeper, but they couldn't give it a name. The first math anxiety measurement scale was created by Richardson and Suinn in 1972. They defined math anxiety as, "feelings of tension and anxiety that interfere with the manipulation of numbers and the solving of mathematical problems in a wide variety of ordinary life and academic situations" (Richardson & Suinn, 1972, p. 551). Not surprisingly, researchers building on their work have found that students with math anxiety report struggling with enjoyment, motivation, and confidence in mathematics (Schillinger et al., 2018). It's no wonder so many of us are struggling to get students to engage in our classroom! If students are struggling with math anxiety it's very difficult for them to enjoy our class, be motivated in math, and have mathematical confidence. If we want students to be excited and ready to learn math, we must help students express their feelings about math and support them in processing their math anxiety as they build a new relationship with math in your classroom. You'll learn practical activities to make this happen in Chapter 3, so keep reading.

In case you're tempted to fluff off math anxiety, you should understand that when students with math anxiety see numbers, hyperactivity takes place in their amygdala, the brain center associated with processing fear (Young et al., 2012). That is important for us to know because when activity in that brain center increases, problem-solving ability decreases. When your students with math anxiety come into your classroom and see a warm up full of problems from last night's homework, it's as if they are walking into a room and seeing something they fear like a lion or a shark.

It's not just problem solving that is impacted when the amygdala experiences hyperactivity. The amygdala is also connected to our primal fight, flight, freeze, or fawn response and when it's overstimulated it causes panic (Young et al., 2012). One of the biggest challenges teachers have with students who struggle is retention. They feel kids were engaged one day and did a great job with the lesson, but when they return just twenty-four hours later it's like the lesson never happened. This is extremely frustrating as math concepts build on each other and it's often vital to have that prior knowledge to move forward. However, when you feel panic, as you do when the amygdala is overstimulated with math anxiety, it's nearly impossible to allow working memory to function properly let alone move information into long term memory. It's not your fault students are struggling with retention in your class, their brains have been hijacked by anxiety. It should be no surprise then that students are struggling with retaining math concepts from day to day when math anxiety is so widespread and present.

Another common challenge many math teachers mention is the struggle with students knowing their basic math facts and being mathematically fluent. We will dig deep into fluency in Chapter 6, but for now we must understand that if we have students with math anxiety, the sight of numbers kicks their anxiety into high gear initiating the "fight or flight" response. So when students are presented with a fluency practice worksheet it's easier to fight, aka check out, trash it, or goof off, than do the work. Again, it's not your fault. The struggle may not be in the way you're teaching basic math facts, the struggle for students may be that their brains are in panic mode and that makes it challenging to problem solve and recall facts accurately. In Chapter 6 we will be talking more in depth about math facts and fluency.

When students experience math anxiety, they are more likely to check out, and less likely to achieve at high levels. When students struggle to achieve at high levels in math and find themselves in remedial math classes or intervention classes it in turn fuels their anxiety telling themselves they're "just not a math person." The cycle continues in this way, contributing to the gatekeeping cycle of mathematics.

> When students have experienced math trauma or experience math anxiety they struggle to succeed in math. This creates a gate, blocking their way to success in mathematics and trapping them in the gatekeeping cycle of mathematics.

Gatekeeping Cycle 2: Teacher Gate

The second gate in the way of improving outcomes for students who struggle is a teacher gate made up of remedial content and teacher expectations.

Teacher Expectations

While there are many ways teachers can show and demonstrate that they either have high expectations or low expectations of their students, we're

GATEKEEPING CYCLE 2

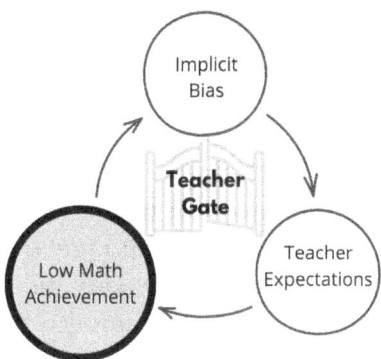

Figure 1.4 Gatekeeping Cycle #2: Teacher Gate – Teacher Expectations And Implicit Bias

going to explore the most impactful way teachers can communicate high expectations to students: teaching grade level content. Researcher John Hattie released a meta-analysis of 252 influences and effect sizes related to student achievement and the third most impactful way to increase student achievement was teacher estimates of achievement (Hattie, 2017). There are a multitude of reasons why math teachers do not give students access to grade level content. For one, many math teachers are bombarded with diagnostic data before we even meet our students. Receiving a report that your entire 5th period class is scoring at a fourth grade level in math makes it all too easy – and subconscious – to lower our expectations of our students and only show them remedial content all year long. I'm going to share my Just in Time Math Intervention Model in Chapter 6 where you'll learn exactly what to do when you find yourself in this situation: You want to have high expectations of your students, but you also live in the reality that they are multiple grade levels behind and have many gaps in mathematical content knowledge needed for success at the grade level content.

This next section took a lot of learning, unlearning, and re-learning for me to write. While I would feel more comfortable leaving the next section out of this book, I cannot. So I'm being vulnerable with you. I may stumble over my words, I may make a mistake that I fix in future editions of this book, but these topics must be part of the conversation about math education. So rather than hide them, push them down deep, and pretend they don't exist or have an impact on students, I'm going to bring them into the light so that we can begin the important conversations around these issues.

Implicit Bias

Implicit bias is defined by the National Education Association (NEA) as "the attitudes or stereotypes that affect our understanding, actions, and decisions in an unconscious manner" (NEA, 2021; paragraph 1). It's important for us to understand implicit bias because it has been shown to impact not only teachers' attitudes towards student behavior, but their expectations of student achievement. Before we go any further I want to stress the importance of understanding that implicit bias is unconscious. We are unaware that we are expressing a bias towards any of our students. That is what makes implicit bias so sneaky and challenging. This section is not meant to bring you any shame, it's meant to make you aware that we all have biases and they may be impacting students in our classrooms without even knowing it. The first step is to become aware that you may have some implicit biases at play in your classroom. The following are some common ways implicit bias might manifest in our classrooms.

Mispronouncing Names

If you find a student's name challenging to pronounce, you may mispronounce that student's name. While it seems harmless, it's actually a microaggression and it's important that we're taking time to ask how students' names are pronounced and then actually pronounce them that way, not half-heartedly saying, "I tried, but it's too difficult" and continuing to mispronounce their name.

Calling On Stereotypically Successful Students More Often

Our implicit bias can steer us to call on students who look like us or look like who we believe will give us the correct answer. We've already explored data showing us that students of color, students with less economic advantage, students with disabilities, and multilingual learners are – and have been for a long time – scoring lower than white students, Asian students, high income students, students without learning differences, and English-only students. Implicit bias may subconsciously steer us to call on students who are stereotypically successful and avoid engaging those who are not because we assume they'll struggle, get the question incorrect, or refuse to answer – potentially kicking off a classroom management nightmare.

Math Problems That Disregard Socioeconomic Status

A common word problem scenario used in many math textbooks around linear functions is the example of skiing. You ski downhill and that's a negative slope. But using this example can be another example of implicit bias especially if you're teaching students who do not have access to such an expensive hobby like skiing. It's not to say you cannot use the skiing example when you discuss slope, but it is to say that you need to spend more time having a discussion about skiing and not just assume that all of your students know what skiing is, how it works, or have experience doing it.

Describing Students With Deficits

While this is more related to deficit thinking, which we explore in the next section, it falls within teachers expectations so I've included it here. I commonly ask teachers in my professional development sessions with schools and districts what their biggest challenges are with their students who struggle in math. I get responses like engagement, perseverance, basic fact mastery, and so on. But I also get statements like:

- ◆ They simply don't sit down and will not stop insulting each other
- ◆ Keeping my low students motivated
- ◆ Students who struggle don't come to class prepared to learn

Any time we're feeling challenged with a student or group of students and start a sentence with "they" followed by something negative, I argue we're participating in deficit thinking. It's subtle. It's a delicate line to cross from expressing challenges to engaging in deficit thinking, but if we're labeling students ("they") by deficits, we are participating in deficit thinking. The danger with deficit labeling is that students pick up on it, and once students pick up on your deficit views of them, it's nearly impossible to help them experience a sense of belonging or achieve academic gains in your classroom.

Lowering Teacher Expectations

In 2018 The New Teacher Project (TNTP) released a research paper titled "The Opportunity Myth". They found some disturbing data related to teacher expectations and student ethnicity. Researchers found that there was a stark difference in the level of assignments teachers give in their classrooms. In classrooms made up of majority white students, 12% of those classrooms contained no grade level assignments. In classrooms that had a majority of students of color, 38% of those classrooms contained no grade level assignments (TNTP, 2018). What makes this data particularly disturbing is that there was almost no difference between students' ability to succeed at grade level content. In other words, students of color and white students succeed at grade level content equally, however students of color are less likely to even have a chance to try it. This may be due to teachers' implicit bias and subconscious thoughts about who can and cannot do math successfully.

Working on implicit bias has been and will always be part of my journey as a white teacher from a privileged background working with majority Black and Brown children at my first school and majority low-income students at all of my schools. I have taken implicit bias surveys and found I still continue to fall prey to implicit bias. It is not something that any one of us is free from, especially if you teach students who are "different" from you. While we can't prevent ourselves from having thoughts rooted in implicit bias, we can put conscious thought to how we react when they come up. Throughout the framework chapters in this book I'll be sharing "check points" of potential implicit bias triggers that I have noticed in my classroom and the classrooms I support as a way to encourage you to be aware of your own implicit biases.

> When teachers subconsciously make instructional decisions based on their implicit biases, usually resulting in lowered expectations of what students can do in mathematics, students struggle to succeed in math. Teachers then inadvertently create a gate, blocking students' way to success in mathematics and trapping them in the gatekeeping cycle of mathematics.

Gatekeeping Cycle 3: Systemic Gate

The third gate in the way of improving outcomes for students who struggle is a systemic gate made up of deficit math labeling and remedial courses. This is big, really big. Systemic inequities are challenging to discuss at best

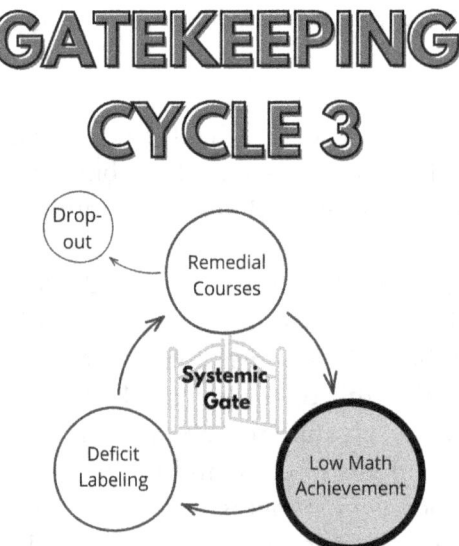

Figure 1.5 Gatekeeping Cycle #3: Systemic Gate – Deficit Math Labeling And Remedial Courses

and divisive at worst, but they must be a part of the gatekeeping and equity conversation in mathematics. The systemic gate is related to institutional racism, a topic I'm not an expert on and is outside the scope of this book, however it cannot be left out. Institutional racism is defined by the editors of *Encyclopaedia Britannica* as follows:

> Institutional racism has been prevalent in American society since colonial times, beginning with its overt expressions in the institution of slavery, Black codes, and Jim Crow segregation… In contrast to the nakedly discriminatory policies and practices of the Jim Crow era, the aspects of contemporary systems and structures that have created social, political, and economic inequities and injustice for Black, Indigenous, Hispanic (Latinx), and Asian Americans are increasingly hidden – ingrained in the standard operating procedures of institutions and eschewing racial terminology.
>
> (*Encyclopaedia Britannica*, 2024, paragraph 3)

Institutional racism is a sad and unfortunate truth of this nation, and has made its way into mathematics education. When we know more, we can do better. When we know and acknowledge that systemic inequities and institutional racism exist in education and mathematics, we can equip ourselves as teachers with awareness and strategies to counteract them. While fixing systemic expectations feels insurmountable, there are things we can do as math teachers within our classrooms to lessen the effects of systemic inequities of the children we have the privilege to teach.

Deficit Math Labeling

Deficit thinking is defined as "the notion that students (particularly those of low income, racial/ethnic minority background) fail in school because such students and their families have internal defects (deficits) that thwart the learning process (for example, limited educability, unmotivated; inadequate family support)" (Valencia, 1997, abstract). Deficit thinking blames the student for behavior and achievement issues instead of acknowledging the role that systemic inequalities play.

While deficit thinking is the choice of an individual, like when a teacher labels students by deficits as discussed in the section prior, deficit math labeling has become commonplace and systemic. This is especially true for how our education system labels students who struggle and have been historically unsuccessful in math. A joint paper from the National Council of Supervisors of Mathematics (NCSM) and TODOS Mathematics for All, gives examples of how ingrained deficit labeling has become with mathematics, most notably that labeling "slow kids," "low kids," "high kids," actually impacts the type of mathematics instruction students receive (NCSM and TODOS, 2016). For example, when teachers are told they are going to be teaching the "low kids" in the intervention period, it's hard for their expectations to not be subconsciously lowered. We have been so ingrained to believe that we need to teach remedial content to "low kids" and help them master the basics before giving them grade level content that math struggles becomes a self-fulfilling prophecy. Students know they are in the "low class" and feel defeated to be repeating the same math class and same math concepts yet again so they disengage and goof off, falling further behind in math, becoming stuck behind the systemic gate and the gatekeeping cycles of mathematics.

Throughout the framework chapters in this book you'll find "deficit thinking connection points" where I've identified situations in my own classroom where I've noticed myself having deficit views of my students or in other classrooms that I support where I noticed the teacher expressing deficit views of their students because this is something we must continually work on within ourselves.

Remedial Courses

The 1997 U.S. Department of Education White Paper I mentioned in the beginning of this section makes it clear: Students need to take high level math courses for long term academic success, it's especially important for low-income students to take high level math courses, and low-income students aren't equally represented in high level math courses (U.S. Department of Education, 1997). This suggests that low-income students are facing a closed gate in mathematics – probably Algebra 1 – and are either placed in remedial courses or run out of time to access the higher level math courses necessary for academic success. This creates a systemic gate for low-income students, making it more challenging to achieve success in mathematics.

Unfortunately it's not just historically underserved students that are experiencing inequities in mathematics course access, students of color are being disproportionately placed in remedial math courses. Data from the US Department of Education, Civil Rights Data Collection Office For Civil Rights released data from 2020 of students who take Algebra 1 in 8th grade (college track) and students who were taking Algebra 1 in 11th and 12th grade (repeating Algebra 1), results are displayed in Figure 1.6. The only ethnic student groups to have a higher percentage of 11th and 12th graders taking Algebra 1 than 8th graders – and therefore clearly repeating the class for the third, fourth, or fifth time – are Latinx and Black students, 10% and 8% respectively. This means a higher proportion of Black and Brown students are stuck in the

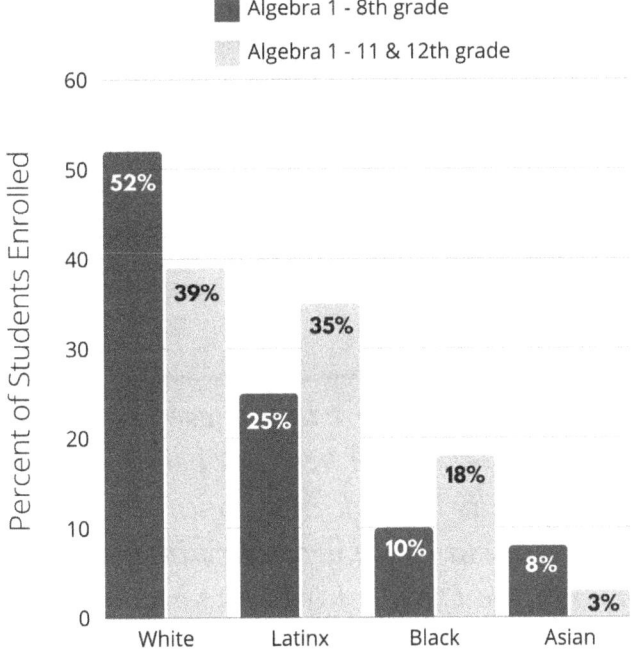

Figure 1.6 Bar Graph Of 8th Grade Students Taking Algebra 1 Vs 11th And 12th Grade Students Taking Algebra 1 By Ethnicity

Created by author from data from U.S. Department of Education Civil Rights Data Collection Office For Civil Rights, 2020.

systemic gatekeeping cycle and Algebra 1 remedial loop than their white and Asian peers.

Being stuck in remedial math courses doesn't just impact the level of math achievement students are able to attain in high school, but it has profound impacts on dropout rates as well. In his journal article, *Solving Our Algebra Problem: Getting All Students through Algebra I to Improve Graduation Rates*, Ron Schachter found that, "eighty percent of high school dropouts cited their inability to pass Algebra I as the primary reason for leaving school" (2013, p. 43). In Figure 1.6 we see that higher rates of Black and Brown students are still taking Algebra 1 in 11th and 12th grade (remedial), so it's sadly no surprise that dropout rates follow suit. The National Center for Education Statistics released dropout rates for 2022 by ethnicity as follows: Asian 1.9%, White 4.3%, Black 5.7%, Hispanic/Latinx 7.9%, and American Indian/Alaska Native 9.9% (National Center for Education Statistics, 2024). When we track students or label students we risk labeling students with deficits which leads to a higher likelihood of being placed in remedial math classes which leads to a higher dropout rate for some, but continued low math achievement for most.

I'll say it again, student demographics are not their mathematical destiny. Not all Black and Brown students will be labeled by deficits or placed in remedial courses. Not all Asian students will be excused from labeling or remedial course placements. As teachers we must be aware that these systemic inequities exist and we must do our best to open these systemic gates for all students by making a commitment to teaching grade level content (with necessary interventions and supports) for any student who has been historically unsuccessful in math.

> *As teachers we must be aware that these systemic inequities exist and we must do our best to open these systemic gates for all students by making a commitment to teaching grade level content (with necessary interventions and supports) for any student who has been historically unsuccessful in math.*

When students experience systemic inequities like deficit labeling and enrollment in remedial courses, they struggle to succeed in math. This creates a gate, blocking their way to success in mathematics and trapping them in the gatekeeping cycle of mathematics.

Students Who Struggle: Stuck Behind Closed Gates

Every year I would ask my students to write and share their Mathography with me (a math autobiography that I share with you in Chapter 3). I always learn interesting things about my students' thoughts and feelings about mathematics, but I remember one student, an 11th grader in my Algebra 1 class, Rudy, told me about a traumatic middle school math experience that created deep math anxiety for him. He enjoyed math all through elementary school, but in 6th grade he was called on to answer a question about one-step equations and got it wrong in front of the entire class. His classmates laughed at him and the teacher responded with a stern "no" and moved on to another student with their hand raised to answer correctly. Now, five years later, he was repeating Algebra 1 for the third time and hadn't passed a math class since 5th grade.

Rudy entered the gatekeeping cycles of mathematics when he first faced a closed teacher gate, as his teacher made him feel stupid and that he wasn't good at mathematics. That traumatic event put up a closed student gate for Rudy. Once he started to fall behind in math due to his math trauma he got flung over to face a closed systemic gate as he repeated and failed, repeated and failed Algebra 1 multiple times. All of it together creates these vicious cycles of gatekeeping that are challenging to free yourself from once you enter. Students who struggle can enter the gatekeeping cycle of mathematics at any one of these individual cycles, but likely get spun around a second or third cycle as well.

Be A Math Gatebreaker

While the gates I've identified paint a bleak picture, there is hope. Inside this book are tools, strategies, and activities you can do with your students to break these gates and become what I call a *gatebreaker*. To be a math gatebreaker is to help any student who struggles with math to recover their mathematical confidence, find academic success, and achieve at high levels in mathematics thereby breaking the gates that have held them back. We know mathematics has long been a gatekeeper, so let's break the gate for our students who need us the most. This book will be your guide.

> *To be a math gatebreaker is to help any student who struggles with math to recover their mathematical confidence, find academic success, and achieve at high levels in mathematics thereby breaking the gates that have held them back.*

✓ Gatebreaker Homework

Each section of this book ends with a "Gatebreaker Homework" assignment, a short task for you to do to prepare to become a gatebreaker. I suggest pausing to complete the homework before you move on to the following chapter. Most of the homework assignments will require you to make a shift in your classroom, however this first assignment is a chance to pause and reflect so that you're ready to become a gatebreaker. You can follow the reflection guide below or download the reflection guide as an editable document in the chapter resources section of www.gatebreakerbook.com. A number of the reflection prompts were inspired by the *Jobs To Be Done Theory* from work by Christensen (2016).

Your "so that" statement

What is your "so that" statement that comes after you tell someone, "I teach math." Some examples could be: I teach math *so that* students find joy in problem solving, I teach math *so that* students can learn skills for high paying jobs, I teach math *so that* I could find a job. What is your "so that" statement?

Juliana's example: I teach math so that I can be a gatebreaker for historically underserved students, helping them break through the gate of Algebra 1 in order to graduate from high school and pursue a life of their dreams.

Explore Your Math Past As A Student

Questions to consider: What is your earliest math memory? What was math like for you in elementary, middle, and high school?

Explore Your Math Past As A Teacher

Questions to consider: Did you want to teach math? Are you teaching the math class you wanted to teach? How is your math class similar or different from the math classes you were in as a student? This is particularly important as we work to increase outcomes for our students who struggle. Why do you teach the way you do and ask yourself if it's effective for all students regardless of ethnicity, economic means, ability status, or how many years they've failed math.

Identify your reason for change

What is taking place that makes you feel you need to change and make some progress differently in your classroom? Why did you pick up this book? (It can be something taking place personally, outside of school, or professionally, within school)

Identifying what you love

What are the things you love most about your classroom/teaching and are essential to keep?

Identify hesitations

Why have you been hesitant to change your teaching practices in the past? Get honest and real with yourself here. What are some fears you have around changing your teaching practices?

Congrats! You've finished all of the prerequisites to learn the B.R.E.A.K. it™ Math Intervention Framework full of strategies, resources, and activities you can start using tomorrow! Thank you for taking the time to get through this part of the book before moving on to the strategies. Without this strong foundation you've built in Chapter 1, the framework will crumble. You're on your way to becoming a gatebreaker! Keep it going.

 ## Gatebreaker Resources

Equity is a topic that I aspire to explore and continue practicing more deeply as I learn from others who are knowledgeable about the topic. I have included a list of additional books and resources that I recommend if you too would like to explore and practice equity more deeply.

Bartell, T. G., Yeh, C., Felton-Koestler, M. D., & Berry III, R. Q. (2022). *Upper elementary mathematics lessons to explore, understand, and respond to social injustice.* Corwin.

Berry III, R. Q., Conway IV, B. M., Lawler, B. R., & Staley, J. W. (2020). *High school mathematics lessons to explore, understand, and respond to social injustice.* Corwin.

Conway IV, B. M., Id-Deen, L., Raygoza, M. C., Ruiz, A., Staley, J. W., & Thanheiser, E. (2022). *Middle school mathematics lessons to explore, understand, and respond to social injustice.* Corwin.

Howard, T. (2024). *Equity now: Justice, repair, and belonging in Schools.* Corwin.

Koestler, C., Ward, J., delRoario Zavala, M., & Bartell, T. G. (2022). *Early elementary mathematics lessons to explore, understand, and respond to social injustice.* Corwin.

National Council of Supervisors of Mathematics and TODOS: Mathematics for All. (2016). Mathematics education through the lens of social justice: Acknowledgment, actions, and accountability (a joint position statement). https://www.todos-math.org/assets/docs2016/2016Enews/3.pospaper16_wtodos_8pp.pdf.

National Education Association. (2021). "Implicit bias, microaggressions, and stereotypes resources" (toolkit). https://www.nea.org/resource-library/implicit-bias-microaggressions-and-stereotypes-resources.

Seda, P., & Brown, K. (2021). *Choosing to see: A framework for equity in the math classroom.* Dave Burgess Consulting, Inc.

Venet, A. (2021). *Equity-centered trauma-informed education.* Routledge.

Reference List

Christensen, C. M., Hall, T., Dillon, K., & Duncan, D. S. (2016). *Competing against luck: The story of innovation and customer choice*. Harper Collins.

Encyclopaedia Britannica. (2024, May 30). institutional racism. Encyclopedia Britannica. https://www.britannica.com/topic/institutional-racism.

Ferlazzo, L. (2022, May 10). Teacher prep should include classroom-culture training. What do you think many teacher-credentialing programs should be teaching that they might not be doing now? *Education Week* (opinion blog). https://www.edweek.org/teaching-learning/opinion-teacher-prep-should-include-classroom-culture-training/2022/05.

Hattie, J. (2017). Hattie Ranking: 252 Influences And Effect Sizes Related To Student Achievement. https://visible-learning.org/hattie-ranking-influences-effect-sizes-learning-achievement/.

Howard, T. (2024). *Equity now: Justice, repair, and belonging in schools*. Corwin.

National Center for Education Statistics. (2023). Mathematics Performance. *Condition of Education*. U.S. Department of Education, Institute of Education Sciences. https://nces.ed.gov/programs/coe/indicator/cnc.

National Center for Education Statistics. (2024). Status Dropout Rates. *Condition of Education*. U.S. Department of Education, Institute of Education Sciences. Retrieved 6/5/24, from https://nces.ed.gov/programs/coe/indicator/coj.

National Council of Supervisors of Mathematics and TODOS: Mathematics for All. (2016). Mathematics education through the lens of social justice: Acknowledgment, actions, and accountability (a joint position statement). https://www.todos-math.org/assets/docs2016/2016Enews/3.pospaper16_wtodos_8pp.pdf.

National Council of Teachers of Mathematics. (2014). Access and equity in mathematics education (position statement). https://www.nctm.org/uploadedFiles/Standards_and_Positions/Position_Statements/Access_and_Equity.pdf.

National Education Association. (2021). Implicit bias, stereotypes, and microaggressions (professional learning). https://www.nea.org/professional-excellence/professional-learning/resources/implicit-bias-stereotypes-microaggressions.

Richardson, F. C., & Suinn, R. M. (1972). The mathematics anxiety rating scale: Psychometric data. *Journal of Counseling Psychology*, 19(6), 551–554. https://doi.org/10.1037/h0033456.

Schachter, R. (2013). Solving our algebra problem: Getting all students through algebra I to improve graduation rates. *District Administration*, 49(5), 43–46.

Schillinger, F. L., Vogel, S. E., Diedrich, J., & Graber, R. H. (2018). Math anxiety, intelligence, and performance in mathematics: Insights from the German adaptation of the abbreviated math anxiety scale (AMAS-G). *Learning and Individual Differences, 61*, 109–119. https://doi.org/10.1016/j.lindif.2017.11.014.

Taylor, M. (2022, April 28). What does fight, flight, freeze, fawn mean? https://www.webmd.com/mental-health/what-does-fight-flight-freeze-fawn-mean.

The New Teacher Project. (2018). The opportunity myth: What students can show us about how school is letting them down – And how to fix it. https://tntp.org/publication/the-opportunity-myth.

U.S. Department of Education. (1997). *Mathematics equals opportunity* (White Paper prepared for U.S. Secretary of Education, Richard Riley, ED 415–119). https://files.eric.ed.gov/fulltext/ED415119.pdf.

U.S. Department of Education, Institute of Education Sciences, National Center for Education Statistics, National Assessment of Educational Progress (NAEP), various years, 1990–2022 Mathematics Assessments. https://www.nationsreportcard.gov/mathematics/nation/achievement/?grade=8.

U.S. Department of Education Civil Rights Data Collection Office for Civil Rights. (2020). Data on equal access to education. Civil Rights Data Collection. https://civilrightsdata.ed.gov.

Vakharia, V. (Host). (2023, May 25). *Math therapy: What even is math trauma?* (audio podcast). https://www.maththerapypodcast.com/.

Valencia, R. R. (1997). *The evolution of deficit thinking: Educational thought and practice*. The Falmer Press/Taylor & Francis.

Young, C. B., Wu, S. S., & Menon, V. (2012). The neurodevelopmental basis of math anxiety. *Psychological Science, 23*(5), 492–501. https://doi.org/10.1177/0956797611429134.

2

B.R.E.A.K. it™ Math Intervention Framework

"I'm tired of fighting the phones"

I was scrolling Instagram dog reels (confession: it's my guilty pleasure) when a DM popped up from Sarah, a veteran twenty-year high school math teacher in Indiana. At the start of second semester she was voluntold to teach the repeater class meaning every single one of her students had failed Algebra 1 the semester prior.

"I am the only one teaching the students who failed math last semester. It's just exhausting and I'm very discouraged. I gave a test and many students failed. I'm so tired of fighting the cell phones," she wrote to me. When I asked her to say more about her frustration she wrote, "I feel as if I'm talking to myself all day. My students don't raise their hands to answer questions and if I ask them a question directly I just get IDK as an answer." I'm wondering if you've felt some of these same feelings because I know I have. The apathy, the disengagement, the struggle to just get started on anything in class. All of these issues made Sarah unhappy to come to work on a daily basis, a profession she loved and had poured her entire career into.

Over the next few weeks she registered for my online PD program and implemented the B.R.E.A.K. it™ Math Intervention Framework in her repeater Algebra class. She watched the videos, she set up her room accordingly, she jumped in, and she saw massive results quickly. Her students started consistently taking notes, answering questions, and turning in assignments. Her test scores improved significantly, but the best thing she feels she's gotten out

DOI: 10.4324/9781003479703-3

of implementing these strategies is, "learning how to engage my students. I have less issues with students disengaging by putting their heads down or playing on their phones." From frustration over phones to phones not even being an issue, that in itself is a win for any math class in our tech obsessed culture.

"My students are truly thriving! I cannot thank you enough! This class was my favorite and I'm actually excited to teach next year," she wrote to me. Within one semester her students had achieved what seemed impossible: an 80% pass rate. She took a classroom full of students who were stuck in the gatekeeping cycle of mathematics and broke it. Sarah became a gatebreaker.

You might be thinking that a transformation like Sarah's is a one off fluke, it's not. Throughout this book you will meet many other teachers who have implemented the strategies and methods with the same insane success. As you implement the framework and strategies revealed in this book I hope you'll be sending me Instagram messages like Sarah's:

"Almost every seat taken, but every student actively working. You SAVED my classroom."

"Absolutely love the collaboration that happens in my room."

The same success is possible for you when you follow the five steps of the B.R.E.A.K. it™ Math Intervention Framework that make up the next five chapters of this book. Let's take a closer look.

B.R.E.A.K. it™ Math Intervention Framework

 Download the framework summary in the chapter resources section at www.gatebreakerbook.com.

The framework is divided into three phases: student engagement, student achievement, and student motivation. There are five total steps that need to be completed in order: build community, routines to boost confidence, engage every student, advance your expectations, and know students' level of understanding. In the chapters that follow I will share practical strategies you can begin using tomorrow to help you through each phase on your way to becoming a gatebreaker for your students.

B.R.E.A.K. it™ Math Intervention Framework
helping teachers break the gatekeeping cycles of mathematics

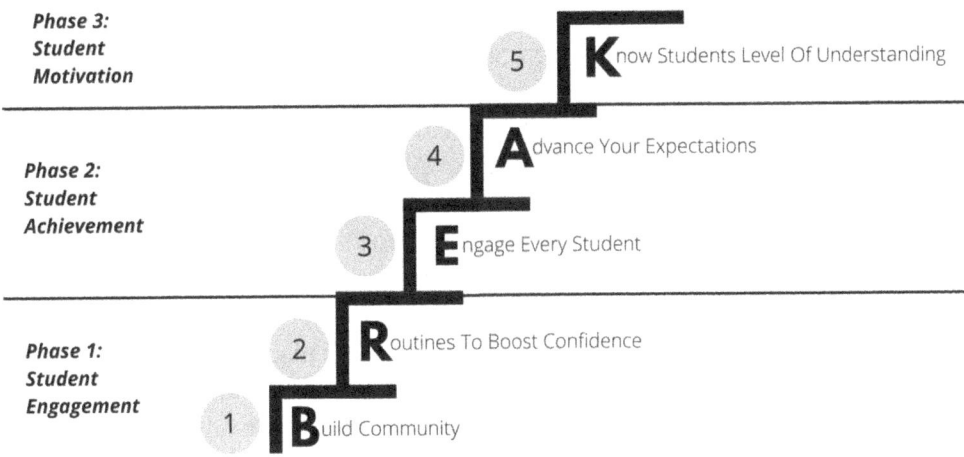

Phase 3:
Student
Motivation

5 **K**now Students Level Of Understanding

Phase 2:
Student
Achievement

4 **A**dvance Your Expectations

3 **E**ngage Every Student

Phase 1:
Student
Engagement

2 **R**outines To Boost Confidence

1 **B**uild Community

Figure 2.1 The B.R.E.A.K. it™ Math Intervention Framework

Phase 1: Student Engagement

Phase 1 has two steps: Step 1, build community, and Step 2, routines to boost confidence. Many of the teachers I support are frustrated by the sound of crickets in their classrooms. They want students to be participating in rich discussion about mathematics, but instead it's like pulling teeth to get students to engage at all. This is where we will start, this is where we must start. To break the first gatekeeping cycle, the student gate of math trauma and math anxiety, we must build intentional community and implement routines that boost student confidence in mathematics. I already know you're tempted to skip the classroom community chapter, but without this foundation, the rest of the framework crumbles. Don't worry, I'm not going to ask you to build a birthday wall or do classroom bingo every unit. Instead, I'm going to give you the rationale and the tools to facilitate a community that is a safe place for students who have been historically unsuccessful in mathematics to thrive. Once we have that strong community foundation it's time for the next step of the framework, routines to boost confidence. In this chapter you'll learn the six engagement structures you can use every single day to get students engaging and talking about mathematics so that you can kiss those crickets goodbye.

Some teachers will need to stay in Phase 1 for a while. If you teach math intervention and you only get to Phase 1 in the first year, that's okay. If all you do is help students feel comfortable in a math class again and finally start participating in the engagement structures, you've won. While you might not see the rise in scores you hoped for, you are setting those students up to have a positive math experience the following year and that is something worth celebrating.

Phase 2: Student Achievement

Phase 2 also has two steps: Step 3, engage every student, and Step 4, advance your expectations. Only once you've completed Steps 1 and 2 should you move on to Steps 3 and 4. In order to break the second gatekeeping cycle, the teacher gate of remedial content and implicit bias, we must learn how to engage every student in every lesson and advance our own expectations of what students can do in our classrooms. Here you'll learn my signature Math Wars Method® so that you know how to deliver content each and every day effectively for students who struggle, boost retention, and no longer feel like you're talking to yourself during your lecture. You'll also learn about the Just In Time Math Intervention Cycle and how to have high grade level expectations of your students even if they have failed math for multiple years and struggle with their basic math facts.

Phase 3: Student Motivation

Phase 3 has just one step: Step 5: know students' level of understanding. This is a phase and step we all desire to get to. We all want to have students who are motivated to show up every day, put in the work, and put effort into tests and quizzes, but just like the other phases, if you don't complete the phases prior, the framework will crumble and you will not experience the gains you hoped for. You might be surprised to hear that in order to increase student motivation you actually need to overhaul your grading and assessment system. In this final step you'll learn my mastery based grading system called the Rethinking Math Assessment Framework™. This might be the most challenging step to implement in your classroom because it goes against everything you've experienced as a student and everything you believe grading should be as a teacher, thus shifting systemic gates in mathematics, but if you really want to be a gatebreaker, you cannot skip this final step.

Let's Become Math Gatebreakers

Once you've implemented all of the steps of the B.R.E.A.K. it™ Math Intervention Framework in your classroom you will officially be a member of the gatebreaker community. Along this journey together you're going to hear not only about my classroom, but about the classrooms of other teachers across the U.S, who have applied the B.R.E.A.K. it™ Math Intervention

Framework in their classroom and achieved massive success for themselves and their students. At the end of every school year I give a survey to the teachers who have learned the framework and ask about their pass rates, engagement, and achievement before and after implementing the strategies they learned. While this data is anecdotal and self-reported, it does provide a glimpse into the effectiveness of the framework. On average the teachers who implement the framework experience 20% more students passing their classes, a 46% increase in student participation, and spend 50% less time planning and prepping for lessons. I know the same is possible for you too. Are you ready to be a gatebreaker? Let's jump in!

⊘ Gatebreaker Homework

- ◆ Understand the B.R.E.A.K. it™ Math Intervention Framework phases and steps
- ◆ Make a plan to read and implement each step of the framework as it is laid out in this book
- ◆ Reflect on the questions below before you begin the framework chapters

What are you most looking forward to learning about in the framework?

Does any of this sound like PD you've received before?

What concerns or hesitations do you have about the framework?

This is about to get really practical! Let's dig in and begin our journey towards becoming gatebreakers. Are you ready? Turn the page and let's get started.

PHASE 1: STUDENT ENGAGEMENT

Phase 1 Step 1
Build Community

Phase 1 Step 2
Routines To Boost Confidence

3

Step 1: Build Community

B.R.E.A.K. it™ Math Intervention Framework
helping teachers break the gatekeeping cycles of mathematics

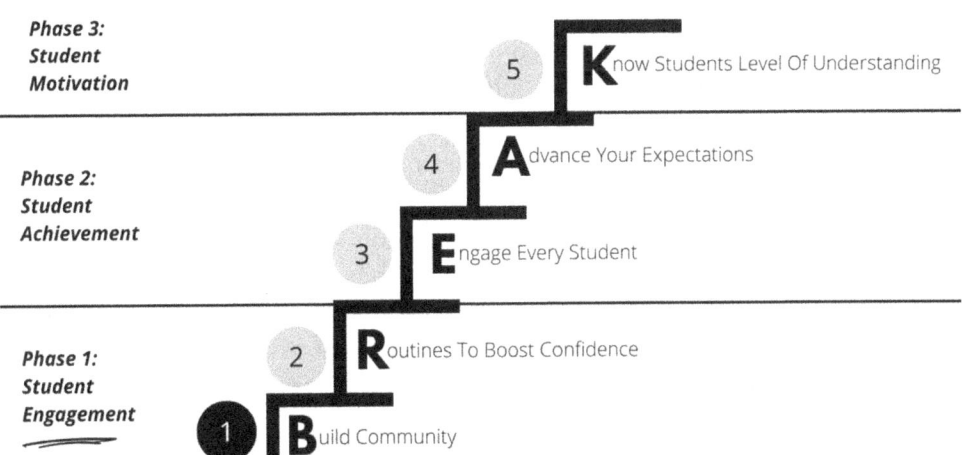

Figure 3.1 Phase 1, Step 1 of the B.R.E.A.K. it™ Math Intervention Framework: Build Community

She Told Me To What?

My first year of teaching was challenging. Really challenging. I was teaching high school Math Support and Algebra 1 in South Los Angeles. I remember one particularly challenging day during 4th period Math Support when I was

DOI: 10.4324/9781003479703-5

doing everything I could to get my students to take the notes and do some independent practice, but despite my best intentions, it was not going very well. I can't remember the exact lesson, but it was pretty much the same every day, students were talking while I was lecturing, getting up to sharpen pencils every few minutes, and constantly asking to go to the bathroom. On this specific day one of my ninth graders, a girl named Justice (pseudonym), was giving me a particularly tough time.

"Justice, take the notes."
"Justice, stop talking."
"Justice, sit down."
"Justice, did you hear what I just said?"

I couldn't help but feel like Justice and I were butting heads more than usual. It was draining and I was incredibly frustrated by the time the bell finally rang.

"Justice, stay after class for a moment, we need to talk" I said. Her response will be etched in my mind forever, "Suck a d***, b****" she told me as she pushed past and left my room. Silence, shock, then tears. My advisory class was next and as I stood at the door trying to hold myself together, our counselors, Ms. Hemmans and Ms. Delgado, saw me trying to hold everything together and asked, "Are you alright?" I couldn't hold it back. Uncontrollable tears streamed from my face. Graciously, the counselors stepped in to cover my class. I ran down the hall into the staff bathroom to cry, trying to avoid making a fool of myself during passing period. Then I heard a knock on the door. It was my principal, she heard I was crying thanks to the phone tree that was the school walkie talkie system. We cried, we laughed, she helped form a plan and a restorative mediation between Justice and me. When Justice returned to class a few days later after the mediation, we were able to have a much better time together for the duration of the school year.

This experience made it glaringly obvious that my classroom community and classroom culture wasn't where it needed to be. As a first year teacher I mistakenly thought it was as simple as having a wall for A+ work to display. I thought it meant being the "cool teacher." I thought playing classroom bingo on the first day of school was enough. But in reality, classroom community is so much more. Over the next few years I would research and try many different activities and approaches to building a positive classroom community and culture. It took me years to fine tune and figure out what students who have been historically unsuccessful in mathematics *need* in a classroom community in order to thrive and achieve the success we all so desperately want for them.

My story about Justice stands in stark contrast to my fifth year in the classroom. I was teaching 9th-grade integrated math 1 and high school math intervention (called Concepts) outside of Denver, Colorado. You'll learn the full story about this class in Chapter 6, but wow had things changed since I had gotten cussed out years earlier. At the end of this particular school year I asked students to make a digital scrapbook, each student completing their own slide in a slide deck about our class. They had five sentence starters they could choose from:

1   My Favorite thing we learned in class was…
2   The thing that I am most proud of doing in this class is…
3   I will remember…
4   My Favorite class memory is…
5   Something that was helpful was…

Some of their responses bring me to tears:

◆   Everyone in class was very helpful if you were ever having trouble with something.
◆   I will remember all of the fun we had learning.
◆   I am most proud of going into Integrated Math 2 and it's amazing because I was in Concepts and I thought that I would be behind in math but now I worked really hard and all the work paid off.
◆   Favorite class memory was getting into groups and working with our other classmates.
◆   I will remember how you let everyone be a part of class.
◆   I will most remember the class vibe and atmosphere.

On top of not getting cussed out or bursting into tears in my classroom this year, my students achieved some massive gains in their math performance. It was no coincidence. Positive community, positive results.

In this chapter I'm going to fast track your success to building a positive classroom community. I know you're tempted to skip this chapter thinking things like, *we have so much math to cover, I'm not going to do "community building activities" that waste time*, or maybe something like, *I don't need to do community building activities I just need to teach them the math.* You can try to start the year jumping into math and putting off this intentional community building like I did my first year, but you know how well that went. I have come to learn that building the intentional community I lay out for you in this chapter is not optional when you teach students who have struggled with and failed math for many years. It is the foundation upon which everything else is built. If you skip this chapter and skip the activities in this chapter, the rest of the framework will crumble and your students will not achieve the success they need in order to exit the gatekeeping cycles of mathematics.

I have come to learn that building the intentional community I lay out for you in this chapter is not optional when you teach students who have struggled with and failed math for many years. It is the foundation upon which everything else is built.

Build Community

We begin in Phase 1, Step 1: Build Community. In this chapter you'll understand why math teachers cannot skip community building and get practical ready to use activities you can use to build the intentional community students who struggle need in order to thrive in your classroom. Classroom community is defined as "a supportive social group in which members feel a sense of belonging and share a common interest, experience, or goals" (Columbia Center for Teaching and Learning, 2021, paragraph 1). Think about the community you've built in your classroom. Do students feel a sense of belonging? Is there a common interest, experience, or goal beyond being in and passing class? How do we cultivate these qualities in our classrooms and does it really matter in math? I don't want you to be stuck in the same ping pong of misbehavior, disrespect, and lack of understanding cycle that myself and countless other teachers have been stuck in because they didn't prioritize intentional community building aimed at helping students who have been historically unsuccessful in mathematics. Without pausing to build an intentional community in your classroom, the gatekeeping cycles of mathematics will continue to turn, keeping students stuck in cycles of low achievement.

Traditional Community Building

When I say "community building activities" what comes to mind? For me, it's class bingo, questionnaires about student interests, decorating an "about me" page to hang in the classroom. As a high school math teacher, those activities seemed like a waste of time. Not only were we losing valuable instructional minutes with these activities, but my older students weren't very engaged with them. Those activities often felt childish or "little kid" as my students would tell me. This childish waste of time feeling is one major reason why math teachers often skip community building and jump straight into content.

Another reason many math teachers skip community building is because building a supportive and positive classroom community takes time and time is one thing math teachers do not have on their side. There is so much pressure on math teachers to catch students up multiple grade levels in a school year and get through their entire textbook that it is all too common for math teachers to skip community building activities and jump straight into content or diagnostic testing.

While get to know you bingo and student interest questionnaires do have value, classroom community is so much more. And when you teach students who have math anxiety and/or math trauma, your approach to classroom community building has specific and unique needs that must be addressed. That is what the rest of this chapter is dedicated to discussing: The unique classroom community needs for classrooms with students who struggle with math.

Essential Elements for Building Effective Classroom Community

An effective mathematics classroom community means that our classroom is a place where students feel safe to speak up, make public math mistakes, and persevere. An effective mathematics classroom community has norms and expectations about what everyone within the community does and says while we're all doing mathematics together. An effective mathematics classroom community is safe for students to make public math mistakes and not be afraid there will be any laughing, snickering, judgment, or put downs. To achieve that kind of classroom community takes intentionality, it will not just form on its own, especially if you teach students who have been checked out of mathematics for years, have math anxiety, or have experienced math trauma. In order to achieve this intentional classroom community we must ensure that students have a sense of belonging in our classroom, feel known in our classroom, and feel safe in our classroom. Without addressing these three elements, our classrooms will not be the community our students need in order to thrive.

> ### ⟲ Math Trauma Connection Point
>
> Research has shown that the problem solving center in the brain as well as recall ability is greatly diminished in students with math anxiety and trauma due to hyperactivity in the amygdala when these students see mathematics. If we want to increase problem solving and recall ability we must take time to build a classroom community where students are able to express their feelings about math, process those feelings, and move forward in a new relationship with mathematics. They must belong, be known, and be safe in our classrooms. If we skip community building the math trauma and math anxiety will only continue to re-emerge and continue making retention and fact recall challenging no matter how many different ways we try to teach it.

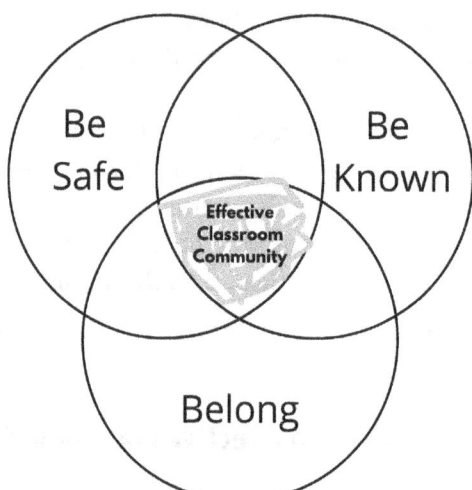

Figure 3.2 Three Essential Elements For Effective Classroom Community Building

Belong

> Belonging is predicated on the idea that no one is left behind, excluded, overlooked, or rendered invisible. Equity is not possible without belonging. Belonging is rooted in the idea that an individual has a persistent feeling of security and support, and that there is a sense of acceptance or membership in a group of community. Belonging tells students that they matter, and truly belonging means that students feel a sense of belonging that is unwavering and not only recognizes but affirms their identities and contributions toward being a part of something bigger than themselves.
>
> (Howard, 2024, p. 19)

The first essential element of effective classroom community is that students must feel that they belong. The above quote from Dr. Tyrone Howard's book, *Equity Now*, says it all. If we want students who have traditionally struggled with math to change direction and begin thriving in our math class, we must ensure students feel they belong in our classrooms. For years some of our students have felt left behind, excluded, and overlooked in their math classes. And when they do, it becomes easier for them to disengage from the class, after all, they often have no connection to their peers or the teacher. For years some of our students have felt invisible in math class, never getting called on, and seldom even being asked how they are doing on a personal level by their math teacher.

Ensuring that students feel a sense of belonging does have unique challenges in our math classrooms. There is the time piece, as math teachers we just don't have time to pause the content to ask how students are doing when administration has asked us to make two or more years of growth with our students. So how do we help math students feel like they belong? First, it's essential that we get to know our students (more on that in the next section) but we also must dig into ensuring students feel accepted and that your acceptance is unwavering and is not influenced by how they perform mathematically or their history of failure in mathematics prior to stepping foot in your classroom. This unwavering acceptance is mentioned in Dr. Howard's definition of belonging and echoed in Alex Venet's work on unconditional positive regard.

In her book, *Equity-Centered Trauma-Informed Education*, Alex Venet defines Unconditional Positive Regard to mean, "I care about you, you have value, you don't have to do anything to prove it to me, and nothing will change my mind" (2021, p. 98). While most educators would agree that they feel this way about students, it becomes increasingly challenging to practice day in and day out when you are met with students who seem to not care about your class, when you see little to no effort being put into the assignment you spent so much time planning, and when you are constantly addressing misbehaviors instead of actually teaching math. I think back to my experience with my student, Justice, who you met at the beginning of this chapter. I could have disregarded her for the rest of the school year and no one would have blamed me after what she said. Or I could have chosen to practice unconditional positive regard, had a restorative mediation, and welcome her back into our classroom, making my unwavering acceptance of her clear not only to Justice, but also to the rest of the class. She belonged.

Another way to encourage belonging in mathematics classes is by developing students' mathematical identity and agency. On the surface it's important that our students see people who look like them doing mathematics in order to develop a mathematical identity. If you teach Black and Brown students, it's essential that you are exposing your students to Black and Brown mathematicians and not only talking about white or Asian mathematicians. But identity goes deeper than that as well. In a blog post for NCTM, Robert Berry defines identity and agency as follows:

> Mathematics identity is defined as how learners see themselves mathematically and how they are seen by others (teachers, parents, and peers) as doers of mathematics. It also refers to a perception of self as a participant in mathematics. Agency is one's identity in action and the presentation of one's identity through participating in mathematics in personally and socially meaningful ways.
>
> (Berry, 2018, paragraph 6)

But here's the challenge: Building a positive mathematical identity means feeling confident that they can do mathematics, building mathematical confidence means feeling comfortable doing mathematics, and comfort doing mathematics in a classroom means feeling like they have a supportive classroom community where they belong, feel known, and feel safe.

Be Known

The second essential element for building an effective classroom community is that students must be known. In classes like Social Studies or Language Arts it seems easier or more natural to weave in conversations about students' lives and histories with the curriculum and content, but it feels more challenging for math teachers. In Dr. Pamela Seda and Dr. Kyndall Brown's book, *Choosing To See*, they share their ICUCARE (pronounced "I see you care") framework for equity in the math classroom. Their book is a must read for any math teacher and the "U" in their framework stands for "understand your students well." If teachers are to create more equitable math classrooms, particularly for Black and Brown students, Drs. Seda and Brown are completely clear that relationships are at the heart of the work. They write,

> Assumptions are usually constructed from stereotypes. Instead of buying into preconceived notions of students based upon their racial and ethnic backgrounds, their grades, or their test scores, teachers need to respect their students enough to take the time to find out who they are as people. Teachers can use a variety of strategies to find out their students' backgrounds and personal interests and then use that information to build a classroom community and design lessons.
>
> (Seda & Brown, 2021, p. 65)

To understand your students well is to know them as people, not as test scores, not as "repeat kids," not as the "low kids," but to know them as human beings. Without pausing to get to know your students on a personal level, the classroom community will not be strong enough to foster a community where students feel safe to take risks and make public math mistakes, students will not feel like they belong in our math classrooms if they do not feel known as people first. Additionally, if you teach students who are different from you (ethnically, racially, socioeconomically, ability, linguistically, etc.) this element of community is of particular importance. The best way to reduce your subconscious implicit bias is to get to know your students. The American Psychological Association (APA) Division 15 released a brief about implicit bias titled, "How White, Middle Class Teachers Can Apply Psychology to Teach Students Who are Different From Them" (Rimm-Kaufman & Thomas, 2021) with practical tips and valuable insight for teachers to consider. Let me just clarify that I am a white teacher who taught mainly Black and Brown students so this is something I need to hear as well. And I'm not alone: Over 80% of teachers in the US are white and 51% of our students identify as students of color and one-fifth of students live in poverty (Rimm-Kaufman & Thomas, 2021). The chances that you're reading this book and teach students who are "different" from you in ethnicity or socio-economic status, is highly likely. Those of us in this situation, myself included, need to be keenly aware of how our implicit biases can manifest in our classrooms and the best way to reduce problems that stem from implicit bias is to learn about our students. One of the suggestions from Dr. Rimm-Kaufman and Dr. Thomas to reduce teachers' implicit bias and connect with students who are different from them is to learn about students and their perspectives. They write, "Learn more about your students so you can understand their perspectives. Take time to understand your classroom from their point of view, identify their strengths and interests, cultivate empathy for them, and appreciate their uniqueness" (Rimm-Kaufman & Thomas, 2021, p. 2).

⟳ Implicit Bias Connection Point

One of the top recommendations for reducing implicit bias is to get to know your students. Get to know them as people, their interests, and their community. This is especially important if you teach students of a different racial, ethnic, socioeconomic, or ability background than yourself. Implicit bias bubbles to the surface when your students are "other" to you. The more you get to know them through community building activities the less "other" they will feel and therefore less likely to subconsciously trigger your implicit bias.

Gatebreaker Tools

So how do we get to know our students as human beings in a meaningful way, respectful of age, and not wasting valuable instructional time? I share my favorite activities in the next section, but some wonderful additional resources can be found in the box below.

ICUCARE Framework (pronounced "I see you care")

What it is: A framework for equity in the math classroom created by Dr. Pamela Seda and Dr. Kyndall Brown, shared in their book, *Choosing To See.*

How it helps you get to know your students: Chapter 3 titled, "Understand Your Students Well" offers valuable insights and tangible activities to build a positive community in your classroom starting with getting to know your students in meaningful ways. I appreciate the constant connection to mathematics offered throughout the book.

More information: Seda, P., & Brown, K. (2021). *Choosing to see: A framework for equity in the math classroom.* Dave Burgess Consulting, Inc.

Kagan Cooperative Structures

What it is: Kagan Cooperative Structures are classroom routines and instructional practices that get students actively and authentically cooperating with their peers.

How it helps you get to know your students: One section of the book on Classbuilding is filled with "get to know you" type activities you can do with your whole class so that you can get to know your students and they can get to know each other. Another section of the book on Teambuilding contains "get to know you" type activities meant to build a strong and safe bond between small group, or team members in class. While some teachers criticize Kagan Structures for being only for younger students, I used these activities consistently in my high school classroom and students truly enjoyed them.

More information: register for a Kagan training or purchase the book: Kagan, M., & Kagan, S. (2015). *Kagan cooperative learning*. Kagan Cooperative Learning.

Intentional Community Building Tools

What it is: Four ready-to-implement activities that will help you get to know students in meaningful ways in order to build a positive community in your classroom.

How it helps you get to know your students: These activities are specifically created for students who have been historically unsuccessful in mathematics to express and overcome math anxiety and math trauma. You will get to know your students and their math pasts more deeply as you help them create a new relationship with mathematics in your classroom.

More information: Keep reading. All four activities are shared in the next section.

Be Safe

The third and final essential element for building an effective classroom community is that students must be safe. There is physical safety, but there is also emotional safety and we will discuss the need for both in order to build a positive classroom community.

Physical Safety

Yes, it's important that your physical classroom is safe and free from harmful objects, but it's also important that students feel that it is safe to be a student in your classroom. Does everyone just get up whenever they feel like it, is it a chaotic environment, are there fights in this classroom, do students steal things from each other when the teacher isn't looking? I must admit, my classroom wasn't always safe. Students would just get up to sharpen a pencil whenever they wanted and most would bonk a friend on the head during their return to their seat. Now I realize how unsafe many students must have felt in my classroom. *What if someone just hits me or touches me,* they must have been thinking. Now I realize the mental space that worry must have been taking up in my students' brains. No wonder they couldn't focus during a lesson or retain information from day to day, their brains were in this constant state of worry and anxiety by just being in my classroom because they did not feel physically safe. It is our job as adults in this situation to ensure our classrooms are physically safe spaces so that students are not in a constant state of stress and/or anxiety. A lower stress environment makes an optimal learning environment and ensures that students have the mental capacity, ability, and desire to learn from you.

One necessity to increase the physical safety in your classroom is to have rules, norms, and consequences for not meeting the expectations. You must enforce the norms and there must be consequences for not meeting them. Structure makes safety possible. Structure allows students to know what to expect in your classroom and your follow through with consequences adds to the safety and structure. Once students know what to expect and that you mean what you say, their affective filter will lower, making learning possible. I suggest a routine like CHAMPS (Sprick et al., 2021) – Conversation level, Help, Activity, Movement, Participation, Success – to first brainstorm your expectations and secondly, communicate them to your students effectively.

Emotional Safety

Feeling safe in our classrooms is also about emotional safety. Emotional safety in math, especially when you teach students with past math trauma, is the most important part of your classroom community building. I'll be sharing tangible activities that help foster emotional safety in mathematics in the next section, but before we get there let's understand the importance of this type of safety. Students walk into our math classroom and immediately wonder:

Is it safe to share my math thinking in this classroom?
Will someone laugh at me when I get an answer wrong?
Will the teacher call me "dumb" or just think I'm dumb?

When students are constantly worried someone is going to laugh at them if they get an answer wrong or if they feel unsure about how the teacher is going to respond to a wrong answer, students' brains do not have the space to also learn new information, problem solve, or recall math facts. All classrooms need to build community, but math intervention classrooms or classrooms with students who have traditionally struggled with math need something additional, they need a community that creates this emotional safety and gives them hope that they belong in this math class, that they *can* be doers of mathematics.

The best way to find out if students feel this type of emotional safety in your classroom is to ask them. I encourage you to give an exit ticket with one question like, *Agree or Disagree: I feel safe to make mistakes publicly in this class,* or, *If someone gets an answer wrong in this class, what happens next?* Then look through the responses. Do your students feel emotionally safe in your classroom?

⌒ Math Trauma Connection Point

Students with math trauma experience hyperactivity in their amygdala when that math trauma is triggered causing their primal brain to panic and enter into flight, flight, freeze, or fawn mode (look back to Chapter 1 for more on this). Implementing trauma informed practices to help students process their math traumas will help calm down their amygdala's so that they can retain information, improve recall ability, and problem solve. Without implementing strategies to calm the amygdala, students will continue to struggle with retention, recall, and problem solving.

Intentional Community Building Tools

Now it's time to make everything we've discussed and make it tangible with ready-to-use community building activities specifically designed to help students process and overcome math anxiety and math trauma so that they can belong, be known, and be safe in our classrooms as they work to achieve math at high levels. I've been in well over a hundred secondary math classrooms over the years as an instructional coach and I can tell within five minutes of being in someone's classroom if they've built the community students who struggle need in order to thrive. I want to fast track your success with students who struggle so I'm laying out all the activities you need to do to achieve that community. The suggestions below are all created with trauma-informed practices in mind, but applied specifically to math.

Download these activities that are ready to print in the chapter resources section at www.gatebreakerbook.com.

Figure 3.3 Community Building Activities to Give Students H.O.P.E. In Mathematics

Honesty: Math Matters Paragraphs

To help students process and move past their math trauma it is vital that we look to trauma-informed research for guidance. While it is not our job as math teachers to process emotional trauma – students need to see a trained counselor for that – it is important to help them uncover and process their math trauma if they are going to achieve different and better results in our math class. The first step of trauma-informed work is relationship building and an essential first step of a relationship building is honesty. How many times have you been asked, "Ay Miss, when am I going to use this in real life?" I know it was one of the most frequently asked questions in my classroom. After all, learning how to factor quadratic functions isn't something you get asked at Starbucks or at the gym a whole lot. Many teachers answer this question by talking about the beauty of math, that math is used in many fields of work, and the joys of problem solving as a skill to be practiced in math. While those things are all true, those reasons may not connect with all students, especially students who have been historically unsuccessful in math for many years. I prefer to answer that age-old math teacher question by being honest about something many of my students feel more passionate about: graduating from high school. When students ask me, "Ay Miss, when am I going to use this in real life?" I give them my honest answer, "You're probably not going to get asked to factor quadratics walking down the street, but the state of [insert your state here] says you have to pass Algebra 1 to get a high school diploma and I want everyone in here to get a high school diploma so we just need to learn it." Once I began answering the question with that type of honesty, students stopped asking because it connected with them on a deeper level. Let's make this a tangible community building activity with an activity I call *Math Matters Paragraphs*.

Activity 1

Activity overview: Students will read and reflect on article excerpts and data about inequities in math scores and math education so that they understand the importance of passing math classes in order to achieve their longer term goals of a high school diploma and college acceptance.

Activity time estimate: 30 minutes

Materials:

- ◆ 8 paragraphs and/or graphs about math inequities. You can use excerpts from chapter 1 in this book or do a google search for additional resources. You can download done for you paragraphs for this activity in the resources section of this chapter at www.gatebreakerbook.com.
- ◆ "Say Mean Matter" graphic organizer for students which can be downloaded with the resources.
  - – Say: Pick a quote from the paragraph
  - – Mean: What does the quote mean to you in your own words?
  - – Matter: Why does this quote matter to you

Teacher prep:

- ◆ Print the eight paragraphs on single sided paper. Pro tip: Use colored card stock and put each paragraph in a page protector sleeve for durability from period to period
- ◆ Print the "Say Mean Matter" graphic organizer, one per student

Directions:

1 Explain the activity to students:
   a We're going to read different article excerpts about why math matters and reflect on the handout that I give you. For each paragraph that you read with your group you're going to fill out the "say" box with a quote that resonates with you. Then you're going to fill out the "mean" box with what that quote means in your own words. Then you're going to fill out the "matter" box with why that quote matters to you. You'll have five minutes to read the paragraph and fill out your notes, then you'll have five minutes to share with your team.
2 Pass out the graphic organizer
3 Place one paragraph at each group

4 Begin the activity

    a Set a visible timer for five minutes to give students time to read and fill out their graphic organizer. This is quiet reading and reflection time.

    b Set a visible timer for five minutes and ask students to share what they wrote. Pro tip: Label your seat numbers (if you have four students in a team, you have seats one through four) and tell students which seat number is to start sharing. This will cut down on wasted time spent deciding and deflecting who starts sharing.

    c Have one student from each team take their paragraph and hand it to the next group (paragraph one goes to team two, paragraph two goes to team three, etc.).

    d Repeat the process until all groups have seen all paragraphs.

Why this activity helps students who struggle with math: Being honest and transparent with students about math inequities may help motivate them to not conform to statistics. Sharing with your students, especially if you teach many historically underserved students, that math has been historically inequitable is a hard truth, but an honest truth. Being honest with your students is a trauma-informed best practice for helping students overcome math trauma in order to build a more positive relationship with mathematics.

Figure 3.4 Students Working Collaboratively On A Community Building Activity

Image courtesy of Sarah Strange.

Activity in photo: Sara Van Der Werk, *100 Numbers Task* (2022).

Ownership: Respect Circles

If you teach students who have experienced trauma, one trauma-informed best practice is to help students feel ownership and agency. We can take this practice and apply it helping students who have experienced math trauma as well. In order to increase students' ownership in our classroom we must provide an opportunity to co-create norms and expectations with our students. I know what you're thinking, *taking time to co-create norms in middle school and high school is unnecessary and a waste of time*, however when done right, it's immensely powerful. To co-create norms and expectations in a way that feels relevant for older students, I suggest framing the conversation around the word *respect*. You must understand what respect looks like, sounds like, and means to your students. As hard as it might be to hear, we cannot demand respect from our students just because we are the teacher. We must earn their respect, and in order to earn it we have to know what it means to them. When your students feel respected, they will be more likely to show you respect as well. You can follow along with the directions below or download the ready to print and use resource in the chapter resource section at www.gatebreak-erbook.com.

Activity 2

Activity overview: Co-create a class definition of what a respectful math classroom looks like and sounds like so that students feel safe to take risks and make public math mistakes.

Activity time estimate: 30 min

Materials:

- ◆ Piece of paper

Directions:

- ◆ Tell students, "I want to know what respect means to you so that we can build a community definition of respect in this classroom. I want to respect you in these ways and I want you to respect me and each other in these ways."
- ◆ On your paper write the word "respect" in the middle and draw a circle around it.
- ◆ On your own, quietly take three minutes to create a new circle for each word or phrase that defines the word respect to you. Be sure to include ideas about a respectful classroom to think of at least three words or phrases.

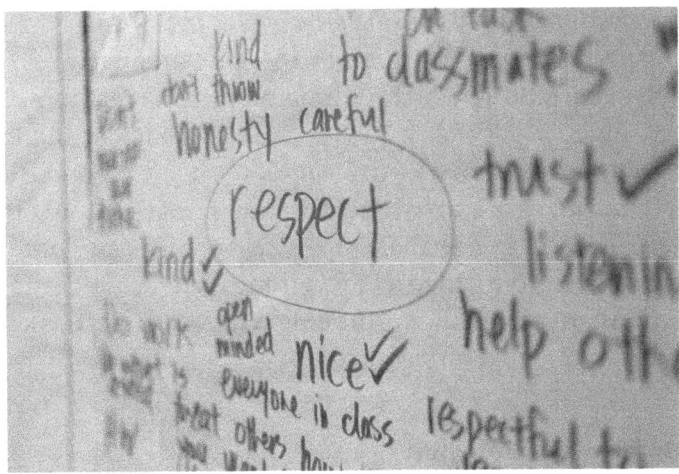

Figure 3.5 Respect Circle Activity
Photography by Author.

- ◆ Seat #1, you will be the scribe for your team. For the next five minutes, share your definition of respect with your team, starting with Seat #2 and going clockwise until everyone has shared.
- ◆ Seat #1, write down at least five words or phrases that your team had in common or that everyone on your team can agree on as a definition of respect and be ready to share out your team summary with the whole class.
- ◆ Ask each team to share their words and phrases of respect while you scribe their responses on a poster you can leave up and reference all year long.

Behavior pro tip: Having this "respect circles" poster in my room was a total lifesaver for classroom management, which seems to always be tricky in math intervention classes no matter how long you've been teaching. Let's take the very common example of students talking over you during a lesson. When you have your co-created definition of respect hanging in your room, you can have the following conversation with your students:

Teacher: *Have I disrespected you today?*
Student: *No*
Teacher: *Right now, you're talking over me and you're talking over your class-mates. I am not feeling respected when you do this. Talking over each other is not in our definition of respect and if I have not done anything to disrespect you, I would appreciate it if you could not disrespect me.*

Why this activity helps students who struggle with math: Collaboratively defining respect in your classroom is a sneaky and age appropriate way to co-create expectations which will help students feel more secure and therefore willing to engage. In order to learn math at high levels, students are going to need to make mistakes, it's just part of the process, but public math mistakes are scary and a common math trauma trigger. Your students will not feel safe to make mistakes in your class until they know how their peers are going to react to a wrong answer. Once the co-creation of expectations are established, students will feel safer to take those risks and make those public math mistakes.

Purpose: Great Wall Of Inspiration

Our students need to connect your class to their larger motivations and sense of purpose in our math class. Another trauma-informed tool teachers can use is to help students see and believe that what they are doing in your classrooms is purposeful. It can be difficult for students who have failed math for many years to find purpose in yet another math class, so we'll need to put in some work to connect mathematics with a larger sense of purpose for our students – especially if they have math anxiety or have experienced math trauma – and this is a perfect activity to bridge that gap. You can follow along with the outline below or get the ready to print activity in the chapter resources section at www.gatebreakerbook.com.

Activity 3

Activity overview: Each student will create a "brick" of what inspires and motivates them to go to school and collectively, the bricks will make a "Great Wall Of Inspiration" to decorate the classroom, help you get to know students and their motivations, and help students get to know each other.

Activity time estimate: 20 minutes

Materials:

- ◆ Index cards or index card sized pieces of scratch paper (even better if it's colored!)

Directions:

1 Display the following prompt as a "Do Now" while your students enter: What inspires you to come to school each day? Is it a person? Your parents or a brother/sister? Is it a potential job? (40 words minimum).

Figure 3.6 Great Wall Of Inspiration In A Classroom
Photography by author.

2   After the "Do Now" is complete, display the following for students: Using your Do Now as a rough draft, choose one person or reason that MOST inspires you and write a final draft (25 word minimum) on the paper I hand to you.
3   Pass out the index card sized papers to students.
4   Allow time to finish (three to five minutes).
5   If your classroom is set up in partners, have them pair-share. If your classroom is set up in groups have each group member share one at a time.
6   Bring the class back together and ask for a few (three to eight) students to share with the whole class.
7   After class, put them up on the classroom wall to form your class "Great Wall Of Inspiration".
8   Keep up your "wall" all year and revisit it with students who are losing motivation!

Pro tip: This can be used as a classroom management tool! If a student is losing focus you can tell them to go take a moment to read the wall of inspiration and help them refocus on why they are here.

Note: This activity usually takes a few turns. Most kids take it seriously and give thoughtful answers. Some students use humor and make funny jokes about a celebrity being their inspiration for school. But sometimes it takes a very serious or somber turn. Students have said that what inspires them is their brother because he's dead, or their cousin because she's in jail. I've also gotten responses that their uncle inspires them because he went to college and he was the first in their family to go to college. There are lots of different things that you will learn about your students as a result of this activity and if a student shares something deeply personal, pull them aside the next day and let them know you read it and you honor it. They want to feel seen.

Why this activity helps students who struggle with math: Doing math day in and day out with our students who have been historically unsuccessful in mathematics can be hard. Students in math intervention classes often lack motivation and desire to persevere in mathematics because they've been disengaged and discouraged for so long. Connecting the daily work of mathematics to a bigger purpose or motivation students have can be a very effective tool to help students overcome math trauma and establish a new relationship with mathematics.

Explain: Mathography

I saved the best for last! I have done this activity every single year as the first homework assignment of the year and it is my absolute favorite activity to do with students. It's called a Mathography and in it, students explain their math biography. There are so many benefits to this activity, but when you have students who have math anxiety, this activity cannot be skipped. There is evidence that writing out the fears and worries students have about mathematics is an effective intervention for math anxiety. Researchers found that when students pause to put their anxieties on paper, the working memory area of the brain that was holding that worry is freed up and can now be used to engage more fully in class (Dowker et al., 2016). This activity allows students to share their math baggage with you in a safe way. You can follow along with the outline below or download the ready to print resource in the chapter resources section at www.gatebreakerbook.com.

Researchers found that when students pause to put their anxieties on paper, the working memory area of the brain that was holding that worry is freed up and can now be used to engage more fully in class (Dowker et al., 2016).

Activity 4

Activity overview: Give students an opportunity to share their personal math story with you so that you can understand their math past, give them validation of their past experiences, and let them know you want them to have a positive experience in this math class regardless of the math past.

Activity time estimate: I like to assign this as a homework assignment during the first week of school and expect it would take ten minutes for students to complete. You can do it during class if you prefer.

Materials:

◆ Printed Mathography prompt or blank paper with the prompt displayed on screen for students to respond to.

Directions:

1 Pass out a paper with the following prompt or display it for students to respond to:

    a Tell me about your experience as a math student from kindergarten until now. What kind of math are you good at? What kind of math do you struggle with? How are you feeling about taking this class?

Pro tip: Give yourself time to read every single Mathography and make written comments before you pass it back to students. Let them know that you read it. They want to know that you read it and that you still accept them no matter what their math pasts look like and that they belong in this classroom. Seeing and reading your validation despite their struggles will help them feel known and feel a sense of belonging in your classroom.

Why this activity helps students who struggle with math: The research is clear: Writing your math fears frees up mental space to build a new relationship with mathematics and overcome math anxiety. Students need an opportunity to share this with you if they are going to achieve at high levels with you.

Figure 3.7 Stock Photo: Wide Angle View Of High School Students

Courtesy of Monkey Business Images and Shutterstock (https://www.shutterstock.com/image-photo/wide-angle-view-high-school-students-1332874982).

My Students And I Were Bored

My classroom was honestly boring, both my students and I were bored. Participation was low with the same couple of kids voluntarily answering or sitting in silence. I felt like I was in an endless cycle of notes and worksheets and it just wasn't engaging and many students were apathetic. I also struggled with classroom management. It's no surprise that pass rates on quizzes were around 80%. For my District Common Finals I had around 60% proficiency, 20% of which were at mastery.

I registered for Juliana's online PD and gradually worked on all of the pieces of the B.R.E.A.K. it™ Math Intervention Framework. Now I always start the semester with a few of the community building activities she suggests to get to know my students and understand their history with math. The Mathography has allowed me to gather history on my students' thoughts and feelings with math and better understand their comfort levels with math. This has been eye-opening as it has shown me that students don't just hate math…they have baggage or trauma related to past experiences in classes that has shown them they "aren't good" or will "never be good" at math. The Respect Circles activity has been super helpful with helping students feel at-ease in my room and reminds them how to treat each other. It has helped me to assist breaking quieter students out of their shells.

Knowing all of this about my students allows me to have proper footing before getting started with content so I can approach my teaching in a way that the students will be receptive to and allows me to build up their confidence and self-esteem in the classroom and hopefully change their view that math is a skill, and because of that, it's something anyone can be good at!

Since I began focusing on community and implementing the framework principles, all students have begun to feel more comfortable sharing and more willing to have small group discussions. Many students have even come out of my class with new friends. Students now approach my class and their peers in a positive and respectful manner, despite what has happened in the past. We start each semester fresh and don't let past experiences influence our present interactions and experiences.

Now there is almost 100% work completion amongst my students. Their confidence and motivation has gone up tremendously. Previously timid students have done exceptionally well, they are willing to participate and share answers, both in small groups and in-front of the whole class. There is also better retention of concepts due to the Math Wars Method® since it allows students to get more practice in a shorter amount of time.

Now I have the lowest failure rate in my department. I have been asked to teach the "support-level" Algebra 1 course due to the implementation of these strategies. My assessment pass rates have gone up to 90% or higher, depending on the class and I have one of the highest mastery rates on district finals. My administrators have been impressed with my ability to gain the respect of my students and achieve success in engaging my students so early in my teaching career.

I even had a team of administrators come observe my teaching a few weeks ago and one of them tracked student engagement every three minutes. I had 96% engagement at every interval. Another tracked student participation and was able to make a replica of my seating chart with names by the end of the lesson because every student was called at least once.

Melissa, Title 1 High School Math Teacher in Michigan

✅ Gatebreaker Homework

Complete the four activities from this chapter with your students:

- ◆ Math matters paragraphs
- ◆ Respect circles
- ◆ Great wall of inspiration
- ◆ Mathography

It doesn't matter what time of the school year it is. The rest of the information in this book will not make an impact on students who struggle unless you start here with these four activities. Do them immediately. Do not wait. Once you do the activities with your students, you're ready for the next step in our B.R.E.A.K. it™ Math Intervention Framework, routines to boost confidence. Get ready to see engagement skyrocket!

Reference List

Berry, R. (2018, August 15). Unpacking identities and agency through the voices of black boys. National Council of Teachers of Mathematics. https://my.nctm.org/blogs/robert-berry/2018/08/15/unpacking-identities-and-agency-through-the-voices#:~:text=Agency%20is%20one's%20identity%20in,through%20their%20words%20and%20actions.

Columbia Center for Teaching and Learning. (2021). Community building in the classroom. Columbia University. https://ctl.columbia.edu/resources-and-technology/teaching-with-technology/teaching-online/community-building/.

Dowker, A., Sarkar, A., & Looi, Y. C. (2016). Mathematics anxiety: What have we learned in 60 years? *Frontiers in Psychology*, 7, Article 508. https://doi.org/10.3389/fpsyg.2016.00508.

Howard, T. (2024). *Equity now: Justice, repair, and belonging in schools*. Corwin.

Kagan, M., & Kagan, S. (2015). *Kagan cooperative learning*. Kagan Cooperative Learning.

Rimm-Kaufman, S. & Thomas, K. (2021). How white, middle class teachers can apply psychology to teach students who are different from them (practice brief), American Psychological Association Division 15. https://apadiv15.org/wp-content/uploads/2021/07/Practice-Brief-Rimm-Kaufman-Thomas.pdf.

Seda, P., & Brown, K. (2021). *Choosing to see: A framework for equity in the math classroom*. Dave Burgess Consulting, Inc.

Sprick, J., Sprick, R., Edwards, J., & Coughlin, C. (2021). *CHAMPS: A proactive & positive approach to classroom management*. Ancora Publishing.

Van Der Werk, S. (2022). 100 number tasks to get students thinking. https://www.saravanderwerf.com/100-numbers-to-get-students-talking/.

Venet, A. (2021). *Equity-centered trauma-informed education*. Routledge.

Step 2: Routines To Boost Confidence

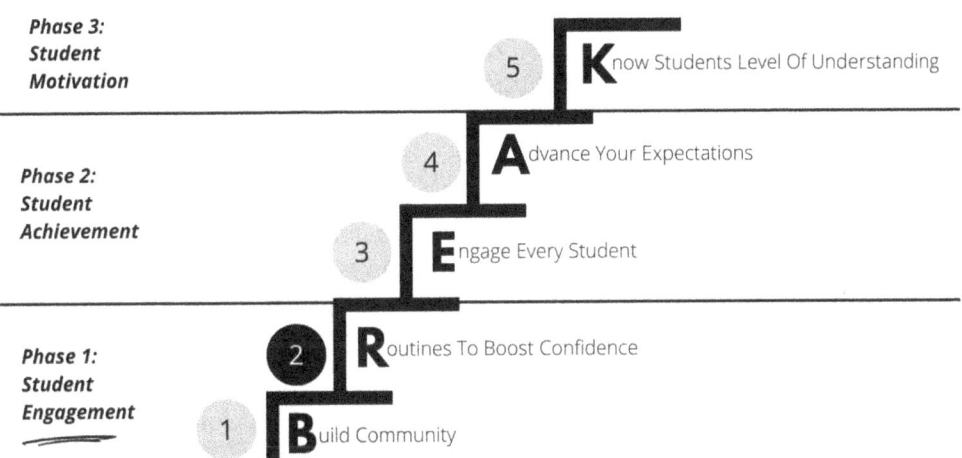

B.R.E.A.K. it™ Math Intervention Framework
helping teachers break the gatekeeping cycles of mathematics

Phase 3:
Student
Motivation

5 **K**now Students Level Of Understanding

Phase 2:
Student
Achievement

4 **A**dvance Your Expectations

3 **E**ngage Every Student

Phase 1:
Student
Engagement

2 **R**outines To Boost Confidence

1 **B**uild Community

Figure 4.1 Phase 1, Step 2 Of The B.R.E.A.K. it™ Math Intervention Framework: Routines To Boost Confidence

How Do You Start Your Math Class?

How do you start your math period? I used to start my class the way I had experienced math class, with homework review. I struggled with engagement and behavior issues from the very moment the bell rang because, if I'm being

DOI: 10.4324/9781003479703-6

honest, very few of my students actually did their homework. So while my intention with starting my class with homework review was to give students a time to ask questions and correct errors to help deepen their understanding, it was just additional time for the majority of my class to goof off and disengage. When I look back at it now, it makes a lot of sense. If you didn't do the homework, why would you care about reviewing the homework? In fact, you might feel excluded, shamed, or embarrassed for not doing the homework. While I certainly didn't mean to create any of those feelings for my students, I now know that's exactly what I did when I spent the first minutes of class excluding the vast majority of my kids with homework review. It got me thinking, how should I start the class period if I want to make sure all students feel included in our community, get in the math mindset, and have a way to actively participate and build confidence?

The first time I saw the "Which One Doesn't Belong?" structure was at an NCTM conference in San Francisco many years ago when I was a district Math TOSA (Teacher On Special Assignment). I walked into the large stadium-style session room and took a seat in the middle. The presenter showed us an image set, gave us the prompt "which one doesn't belong," and asked us to turn and talk with a partner. I was shy and no one was sitting near me so I just sat there thinking to myself that this was such a basic activity. In my mind there was one glaringly correct answer to this question. The presenter quieted the group and began facilitating the structure. I sat there amazed by what other people were saying. Everyone had so many cool reasons for which one didn't belong in their mind, things I would have never noticed on my own. It seemed no one had the same answer as me and for some reason that gave me the courage to raise my hand and contribute my reasoning. I was sold on this structure for getting all students, even the quiet shy ones, even the typically disengaged ones, to participate. When I returned to teaching my own math intervention class a few years later I excitedly tried it out with my own high school math intervention students as a warm up each week.

It was a total transformation of the start of class. Instead of coming in and sitting through fifteen minutes of homework review, students came in, saw the "Which One Doesn't Belong?" prompt on the board, and were immediately excited to be in math class.

Instead of walking in with an immediate invitation to disengage (what else is a kid to do that didn't do the homework?) or worse, an overwhelming amount of problems on a warm up that kick math anxiety into high gear, students walk in and immediately have something to engage with, something to think deeply about, something they feel is an invitation to actively participate no matter how many days of school they had missed, how many grade levels below they were, or whether or not they had done the homework. It changed the entire feel of class not just for the first five to ten minutes, but for the entire period.

Over the years I've come to learn "Which One Doesn't Belong?" isn't the only structure that invites this kind of active participation, engagement, and confidence building. Over the years I've tested dozens of so-called "engagement structures" in my math intervention classes and I've identified six of my all-time favorites that help boost students' confidence in mathematics. When I began sharing these six structures with other teachers, they too were astonished at the engagement it created and loved how students were able to begin the class with a quick math win instead of passively listening to homework review or getting overwhelmed with too many warm up problems. The best part is that they take five to ten minutes of class time and can be used in connection with the content of the lessons, as spiral review, or a way to activate prior knowledge in a not threatening way. Let's dig into these engagement and confidence boosting structures now.

Routines To Boost Confidence

In this chapter we move along our framework into Phase 1, Step 2: Routines To Boost Confidence. Here you'll learn six specific engagement structures that can be used as a quick five-minute warm up with any math content that boosts students mathematical confidence. Think of them as six tools for your student engagement toolbox. I know how it feels to have a room full of crickets, or worse, constant off-task talking, and these six structures will help you get every single student engaging in mathematical discourse.

To help our students who struggle to start achieving at higher rates in our classroom, we must start with the first step of the B.R.E.A.K. it™ Math Intervention Framework: Build Community. Before students will engage mathematically, they must feel the classroom community is a safe place to share their thinking and make public math mistakes. We discussed the importance of community and exactly how to build it in the previous chapter so go back and read it before you read and implement this one. Without the foundation of community, the rest of the framework will crumble. Only once you've built the foundation for a strong community by using the activities in the previous chapter is it time to get students to build their confidence in mathematics. The best way to get students who struggle to grow in their mathematical confidence is by using open-ended structures that focus on reasoning, can be introduced without math, and have multiple correct answers.

Focus On Reasoning

Activity structures that focus on having students explain their thinking instead of calculating one answer are much more approachable for students who struggle. This not only helps you understand students' thinking more, but it provides you with an opportunity to validate their thinking, an integral part of helping students become more confident mathematicians. Research has shown that when math teachers value the different ways students think and reason about mathematics, math anxiety can actually disappear (Boaler, 2019). It is up to us to choose activities and prompts that allow students to explain their reasoning so that we can value their reasoning, heal their math anxiety, and boost their mathematical confidence.

Introduce Without Math

If we want students who have been historically unsuccessful in mathematics to become more confident in our classrooms, we have to introduce activity structures without math first and then layer in the mathematics. Students with math anxiety and math trauma walk into our classroom in a heightened anxious state which diminishes the problem solving centers in their brain. One way we can help calm those nerves and help them activate the working memory area of their brain is to introduce an activity structure without math first. When we introduce an activity structure to students for the first time without mathematics, we're doing all we can to set our students up for being able and willing to participate and grow in their confidence in our classrooms.

Multiple Correct Answers

Providing activities with multiple correct answers is essential to get students to share their thinking confidently. I know you're saying, *but it's math! There is only one right answer for everything!* But that is not totally true. When we make the questions about students' reasoning, there are infinite correct answers because each student might reason slightly differently. Questions and prompts that have multiple correct answers provide us with the perfect opportunity to hear from all students, validate all students, and find success with all students.

When we make the questions about students' reasoning, there are infinite correct answers because each student might reason slightly differently.

Math Anxiety Connection Point

Research has shown that when we give students an opportunity to create a new, positive relationship with mathematics by focusing on valuing the different ways students think and reason, math anxiety disappears (Boaler, 2019). It's up to us to choose activities that will allow students to share their thinking and reasoning so that we can help them heal their math anxiety. These Six Core Engagement Structures do just that!

Six Core Engagement Structures

All of the Six Core Engagement Structures are intentionally chosen to meet all of the requirements described in the section prior: focus on reasoning, can be introduced without math, and have multiple correct answers. Let's dive deep into each one of them so that you can begin implementing them tomorrow and help all students grow in their mathematical confidence and experience a quick math win in your classroom.

You can access done for you resources for each of these activity structures in the chapter resources section at www.gatebreakerbook.com.

Figure 4.2 Six Core Engagement Structures Are Routines To Boost Student Confidence In Math

Which One Doesn't Belong (WODB)

Of all the Six Core Engagement Structures, this one is my absolute favorite. So imagine my delight when, in the midst of coloring yet another princess book with my three-year-old daughter at our kitchen table littered with Crayola markers, she looked up and asked, "Mommy, what does this say?" To my surprise it was a Princess version of Which One Doesn't Belong! I explained the activity to her and we jumped into a fun discussion – because three-year-olds really do say the darndest things – about which image didn't belong in the set. First she told me one reason then I asked her for another and another and another. She loved that every reason was correct, it motivated her to keep going, to keep thinking. My math teacher brain was thrilled! Not only is this structure apparently great for Princess coloring books for preschoolers, it's a fantastic way to get middle and high school students talking about math. Many curricula are beginning to build the Which One Doesn't Belong (WODB) structure into their warm ups and I am here for it! If you're familiar with the structure I'm going to be sharing helpful tips for making the most of WODB for your students who struggle, so don't skip this section. If you're brand new to WODB, prepare to have your mind blown.

Example WODB Prompt: *Using the image below, which one doesn't belong? Justify your reasoning.*

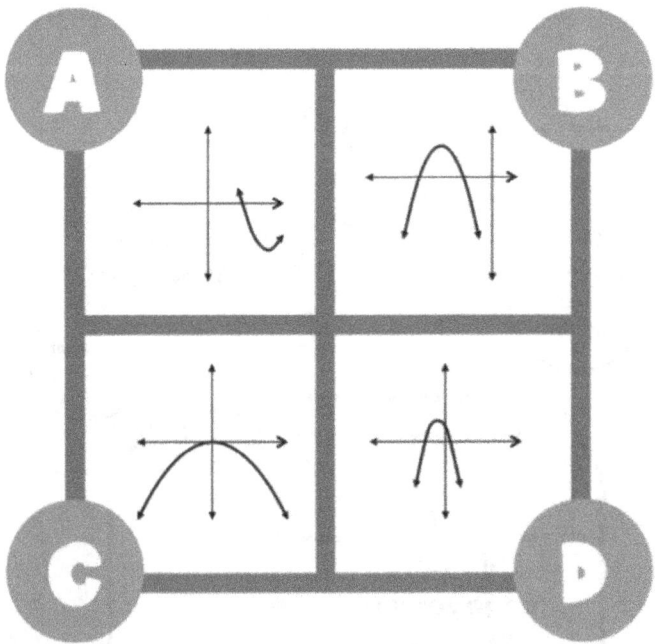

Figure 4.3 Example WODB Prompt With Parabolas
Image created by author.

Activity Structure:

- Display the image set to students with the prompt, "Which One Doesn't Belong?"
  - Give five minutes of think/write time
  - You can ask students to pick why ONE doesn't belong, or you could ask them to come up with a reason why EACH one doesn't belong (that's my personal favorite so that I have more time to take attendance)
- Provide sentence stems for students, "*A doesn't belong because …*" "*ˆ doesn't belong because…*" etc.
- Ask students, "Who wants to tell us why one of these doesn't belong?"
- Scribe the student reasoning on the board
- KEY FACILITATION POINT: Ask, "does anyone have another reason why that same one doesn't belong?" and scribe any reasoning on the board
  - Asking this question instead of just moving on to the next image will really help students see that there are many correct answers!
- Be sure to celebrate the reasoning of all students! Especially if you get a student participating who usually doesn't participate!
- Once you've gotten a few responses for that first image, ask students, "Who has another reason why one of the others doesn't belong?"
- Scribe the student reasoning on the board and repeat the process until you've gotten multiple reasons for each image

The key idea with which one doesn't belong is that it has an entry point for all students and is an activity with multiple correct answers, making it a very safe activity to participate in. The point of this activity structure is to get students to see that there are multiple correct answers and that their answer is valued. The first time you facilitate WODB with your students it might be a little slow and quiet, like it was for me in that conference session, but they are going to hear what their peers say and they are going to hear you validating every single student and that will help them feel incredibly empowered to participate in your class during this activity and beyond. The following are some tips to help.

The following are Which One Doesn't Belong tips.

 Tip #1: Introduce Without Math

It is vital that you introduce this structure without math. Choose a set of four images that have nothing to do with math in order to keep math anxiety low and avoid triggering math trauma. Here's an example:

Figure 4.4 Example WODB Prompt With Flowers
Image created by author.

Tip #2: Add Labels

You'll notice that I've added the labels A, B, C, and D. This helps make the activity more accessible for multilingual learners and I encourage you to label your image sets as well. Instead of having to say, "the upper right doesn't belong because…" students can just say, "B doesn't belong because…" and it helps eliminate a barrier to participation.

Tip #3: Change The Prompt

Many teachers do not like the prompt, "which one doesn't belong?" because *belonging* is so important to so many students and talking about not belonging can be triggering. Many teachers use the prompt, "which one is different?" and I offer that to you as well. Another wonderful prompt I heard a teacher use is "which two belong together?" This really ups the rigor of the activity and forces students to draw deeper connections with the images and mathematics.

Tip #4: Celebrate And Validate

If we want students who are not used to participating in math class to actively participate in our math class, we must validate and celebrate the reasoning that they share. By scribing everyone's responses and giving students verbal validation of their reasoning you will be helping students to feel more confident and comfortable sharing their reasoning all period long. It doesn't matter if the reason is a little superficial or just for laughs, scribe it and validate it.

Tip #5: Add A Word Wall

Using WODB at the beginning of a unit is a wonderful way to assess prior knowledge of a topic in a non-threatening, non-traditional way because it allows students to share their reasoning in a casual way with informal language. You can use WODB at the end of a unit and encourage students to use more academic language in their reasoning to see how far they've come. Adding a word wall with the mathematical vocabulary you want them to use is a great accommodation for students with IEPs and a way to support all students' mathematical language development.

Tip #6: Caution

WODB is not a visual multiple choice problem! I've seen many teachers use this structure in their classroom, but set it up in a way that has only one correct answer. That is not how to use this structure effectively for students who struggle. In order for this to be an effective structure to boost students confidence, there must be multiple reasons why each image doesn't belong in the set and you must be able to validate any answers students give.

Notice And Wonder

This is another popular structure and it helps students to engage deeply with mathematical concepts by encouraging them to share what they observe (notice) as well as what piques their curiosity (wonder) to create powerful mathematical discourse. This is my absolute favorite word problem strategy and while notice and wonder is a very popular structure, most teachers are not utilizing the structure to its fullest capacity. It takes a slight tweak to make this an insanely powerful tool for word problems and tasks that I'm going to share with you shortly. If you're unfamiliar with the notice and wonder structure, you give students an image, graph, or prompt and simply ask, "what do you notice and what do you wonder?" Like all of the Six Core Engagement Structures, it's simple, but incredibly powerful. If we want students to have a more positive relationship with math, confidently conquer word problems, and deepen their conceptual understanding, notice and wonder is going to be your best friend.

I don't know about you, but this is what used to happen in my classroom when we worked on word problems:

Pass out a worksheet with word problems
Crickets, off-task behavior, hands immediately in the air
Go student to student answering questions and mitigating off-task behavior
Answer the same question over and over and over, "Ay Miss, what are we supposed to do?"

At that point I would do what any good math teacher would do and ask, "Have you read the problem?" The answer was always, no. Always.

There is no doubt about it, students are overwhelmed by word problems, yet they are increasingly popular in our math curricula and state tests. Add in students who are multiple grade levels behind and multilingual learners and it makes word problems and tasks feel almost insurmountable. Not anymore.

Once I started facilitating word problems and tasks with the Notice and Wonder strategy, I never had to go student to student to student answering that same dreaded question over and over and over again trying not to show how annoyed I was. Once I started using this structure I had complete confidence that every single student had read the problem, re-read the problem, and understood the context completely. I also had complete confidence that no students had gone ahead and were already shouting the answer to the problem, leaving those who were struggling to understand the problem in the dust feeling "dumb" because they had no clue what to do while other "smart" students were done.

⌒ Implicit Bias Connection Point

To help our students understand word problems, we have to have a deeper understanding of their life, interests, and background. If you are teaching students from a different racial, ethnic, or socioeconomic background than you, it's vital that you take additional time to explore contexts of word problems with your students to make sure they understand the context fully. Often, a context we assume students are familiar with is actually a context they are not familiar with. Assumptions can bring our implicit biases to the surface and impact our students' experience.

Here is where most teachers get this structure wrong. To facilitate notice and wonder effectively for students who struggle, you have to remove the question. Sorry for some tough love, but if you're using a preprinted worksheet or curriculum workbook that has the notice and wonder structure built into the curriculum, you're not using this structure to its maximum potential. You must delete the mathematical question before giving students the prompt, "What do you notice and what do you wonder?" If they can see the question, this structure loses its power. When we remove the mathematical question, students are left with just the facts that they must analyze, observe, and question. Students who "get it" cannot go on, leaving their struggling peers in the dust only to feel like they don't belong in your classroom. Students who struggle get to start on an equal playing field and that will do wonders to boost their confidence and feelings of belonging in your classroom. By removing the question you are also encouraging mathematical thinking. You want students to think of what the question from the textbook might be so that they feel more interested in solving the problem and persevering through any challenges. So how does it work? It's simple.

Sorry for some tough love, but if you're using a preprinted worksheet or curriculum workbook that has the notice and wonder structure built into the curriculum, you're not using this structure to its maximum potential.

Example textbook word problem or task: *You're in charge of buying the chips and soda for a party, you have $55 to spend, the chips cost $9 per bag and the soda costs $2 for a two liter bottle. You want to buy 17 items. How many chips and how many sodas can you buy? Write a system of equations and solve.*

What to show students:

> You're in charge of buying the chips and soda for a party, you have $55 to spend, the chips cost $9 per bag and the soda costs $2 for a two liter bottle. You want to buy 17 items.
> Write down three things you notice and three things you wonder.

Activity structure:

- ◆ Create a T-Chart on the board labeled "Notice" and "Wonder"
- ◆ Present the prompt, graph, or image to students and ask, "What do you notice? What do you wonder? Write down three things you notice and three things you wonder"
  - – Sometimes I like to use the word "Know" instead of "Notice" so feel free to do the same

- ◆ Allow two to five minutes of think/write time
- ◆ Start by asking for students to share one thing they notice or know and scribe everything under the "Notice" column on the board
- ◆ Then ask students to share what they wonder and scribe everything under the "Wonder" column on the board
- ◆ KEY FACILITATION POINT: Have fun! Encourage funny "wonderings," especially when they come from students who don't often participate
- ◆ Now reveal the question and ask students to solve the problem. Be sure to point out if any student "wondered" the same question that was used in the textbook

The following are Notice and Wonder tips.

Tip #1: Introduce Without Math

This is the power of every single one of these Six Core Engagement Structures: You must introduce the structure without math so that students don't feel overwhelmed or anxious. The first time you facilitate notice and wonder, start with a really interesting, visually rich image. You could use an image from a current event, a picture from a trip you went on to share about your life, or an image that shows a hobby you have like cooking, attending sport events, etc. The key is choosing an image that would create lots of discussion. Here is an example:

Figure 4.5 Stock photo: Fans Clapping Hands In Stands

Courtesy of Jacob Lund and Shutterstock (https://www.shutterstock.com/image-photo/sports-fans-clapping-hands-stands-falling-1650156514).

Tip #2: Multilingual Learners

Word problems are especially overwhelming for multilingual learners in our classroom. Not only are there lots of words in English in their math class, there are contexts in word problems that may not be familiar to students who haven't grown up in the U.S. Using the notice and wonder structure for each word problem or task will help multilingual learners understand the context more fully.

Tip #3: Culturally Relevant

Word problems and tasks are also not necessarily culturally relevant for all students. Consider images that connect with your students' cultures to kick off this structure. If a word problem or task isn't particularly culturally relevant, completing a quick notice and wonder as an introduction will help all students understand the context more deeply and provide students with a safe space to ask questions about the context to ensure understanding. For more on culturally relevant math tasks, I'd recommend the book, *Engaging in Culturally Relevant Math Tasks 6–12*, by Matthews, Jones, & Parker (2022).

Tip #4: Validate All Answers, No Matter How Funny

Be ready to have a blast when you facilitate this structure with your students. Your class clowns are going to love telling you what they wonder and getting a giggle from their peers. Celebrate and validate it! Don't dismiss them and say it's off task. Celebrate that they are wanting to participate! For example, using the example math prompt from above, students might wonder:

- How many people are coming?
- Why no hot dogs at this party?
- Is this a school party or a house party?
- Am I just buying one kind of soda? Or am I buying Dr. Pepper and Coke and Sprite?
- Is there tax?
- What store are you shopping at?

While these "wonderings" are funny, they are totally on task. I know that the student who contributed that answer has read the prompt multiple times to come up with these questions. You do want to try to steer some of the "wonderings" to be mathematical, so you might need to remind them of that if it's getting out of hand, but don't forget to have fun with this activity structure!

Tip #5: Don't Use A Workbook

I cannot stress this enough. If you're using a preprinted workbook you're not using this structure effectively because the mathematical question is already there for your students to see. Instead pass out a blank piece of paper and just post the prompt (without the question) on the board. Ask students to write what they notice and wonder on the paper and only once that activity is complete, ask them to open their workbooks and solve the problem mathematically.

Same And Different

Remember the *Highlights Magazine* that seemed to be in every pediatric dentist waiting room? The Same and Different structure seemed to be in most of them. You'd be presented with two similar images, but when you looked closely, they had small differences. One picture has four bananas, the other only has two. The monkey is wearing a blue hat in one image, but a red hat in the other. Don't let the fact that this structure reminds you of your childhood dentist dissuade you from using it with middle school and high school students for math! With this structure you give students two images to compare and contrast with the prompt, "What is the same and what is different about these images?" I like to ask students to find three things that are the same and three things that are different so that they think about the prompt more critically and if I'm using this as a warm up, I get a little more time to take attendance. Same and different is a wonderful way to get students to think about a concept multiple times and in multiple ways and is extremely approachable because there are so many different answers since it's focused on students explaining their reasoning.

Example prompt: *Using the image below, write down three things that are the same and three things that are different.*

$$y = \frac{1}{2}x + 3$$

$$y = \frac{1}{3}x + 2$$

Figure 4.6 Example Same And Different Prompt With Linear Functions
Image created by author.

Activity structure:

- Create a T-chart on the board labeled "Same" and "Different"
- Show the image to students and ask them, "What is the same? What is different about these images?"
- Allow for two to five minutes of individual think/write time
- Ask a student to start by sharing what they see as the same. Scribe their reasoning
- Collect more responses for the "same column" to show students there are multiple correct answers
- Once you have a few responses for what is the same, ask students to share what is different and follow the same process, gathering multiple answers for what is different

The following are Same And Different tips.

 Tip #1: Introduce Without Math

Just like all of the other Six Core Engagement Structures you must introduce this structure without math first and just let students get comfortable participating. It's easy to bring in seasonal holiday image sets with an activity structure like this and students will love it too.

Figure 4.7 Example Same And Different Prompt With Chocolates
Image created by author.

Tip #2: Celebrate And Validate

I encourage you to make a big deal about everyone who participates that might not normally participate. This is a great activity to make every single student feel comfortable participating and to validate their unique thinking.

Tip #3: Add A Word Wall

As with WODB, Same And Different can be a wonderful way to get students using academic vocabulary. If students are struggling with the mathematical words and definitions, consider a word wall to help guide students to the words you'd like them to try using to explain their reasoning. This is particularly helpful for multilingual learners and students with IEP goals around academic language use.

Would You Rather

You're likely familiar with the Would You Rather game. I'm talking about the game you might have played in grade school. It probably went something like, "would you rather go to the movies with [insert name of the most popular guy at school] or kiss [insert name of another popular guy at school]?" Well, this game is making an appropriate comeback in math class! This structure is a great way to get students talking and help them feel comfortable explaining their thinking with a prompt that encourages them to make a choice and take a stand. You will also learn a lot about your students and their interests with this structure, even when you use it with math. One of the things I love most about the Would You Rather structure is that it is really helpful for thinking about math in real life. Just like the other structures, there are infinite correct answers to questions posed with the Would You Rather starter and it's all about how students explain their reasoning, a real challenge to get students to do in mathematics. The purpose of this activity is to get students comfortable making a choice and explaining and justifying their reasoning in a safe way. When students practice explaining their reasoning in the context of these warm ups, they will feel more confident explaining their mathematical thinking when needed.

Example prompt:

> Would you rather buy 8 oz of shredded cheddar cheese for $2.50 or 8 oz of sliced cheddar cheese for $3.50?

Activity structure:

- Show the images or prompt to your students and ask them, "Which would you rather do/have and why?"
- Allow for two to five minutes of individual think/write time
- Ask one student to begin by sharing their choice and their reasoning & scribe their reasoning on the board
- Ask your class, "Did anyone make that same choice, but for a different reason?" Call on students and scribe their reasoning
- Once you have a few reasons for that choice, move to the other choice and ask students to justify their reasoning. Scribe their reasoning

The following are Would You Rather tips.

 Tip #1: Introduce Without Math

The goal of these Core Engagement Structures is to get everyone actively participating and keep math anxiety low. It's essential that you introduce this structure without math before you layer the math in. There are many fun ways to use this structure without math, here's a great intro prompt:

> Which topping would you rather have on your hot chocolate, whipped cream or marshmallows?

Tip #2: Make It Kinesthetic

Ask students to pick a side of the room. If you think it's option A stand on the left, if you think it's option B stand on the right. This is especially helpful if you're teaching a block period and want to give a little movement and brain break in the middle of the period.

Tip #3: Bring The Real World In

Take a picture while you're grocery shopping and comparing two different products to buy. Your mind automatically begins calculating, but you can help students see that math is everywhere and can be used to help in decision making. Another great real world prompt is comparing phone plans or gym plans. Give your students the facts and ask them which plan they'd choose.

Tip #4: Validate All Reasoning

You will be amazed by how much you learn about your students' personal lives with this activity structure. If you give them two phone plans and ask would you rather have phone plan A or phone plan B, they may not always choose the lowest price. They may say things like, "I'd choose plan A because I know plan B gets bad service in my neighborhood." Sometimes it's "I'd choose plan A because it's the lowest cost to start and I don't get paid until next week." Make sure you validate all reasoning and not just what is the most cost effective or obvious to you from a math standpoint. Honor their reasoning even if it's different from yours.

How Many

I first learned about this structure when a colleague told me about Christopher Danielson's (2019) book, *How Many?*. I highly encourage you to get the book and learn about this structure. Even though this structure may only seem fitting for elementary aged students, I have had many high school teachers tell me this is their favorite Six Core Engagement Structure and that their 9–12th graders love it. The premise is simple: give students an image and ask them to create as many questions about the image that begin with the stem, "How many." The key is choosing an image that will get students talking. I love choosing images about cooking, a city street, a current event, or something similar. You can also tie in the current unit or get students thinking about the context of a task from the unit with this prompt. For example, if you're teaching a unit on area or volume you can pull in images with shapes you'll be discussing in the lesson. If you're introducing a task to students about starting a dog-walking business you can show a dog-walker image with this structure. It's very engaging for older students despite many dismissing it, thinking it was made only for younger students.

Example prompt: *Give students a visually rich image and ask them to create as many questions about the image that begin with "How Many?"*

Activity structure:

- ◆ Show the image to students and give them either of the following prompts:
    - – Create as many questions as you can about the image that begin with, "How many…?"
    - – Create as many mathematical questions about the image as you can. Sentence stem ideas include: How many… How much… What is the (area, perimeter, volume, measure, etc.)…
- ◆ Allow for two to five minutes of individual think/write time
- ◆ Ask students to share their questions. Scribe their reasoning

Figure 4.8 Stock Photo: Red Onion And Spices Isolated On White Background

Courtesy of MaraZe and Shutterstock (https://www.shutterstock.com/image-photo/red-onion-spices-isolated-on-white-585422516).

The following are How Many tips.

Tip #1: Make It Seasonal

This is a great prompt to use with holidays! You can find images of a Thanksgiving dinner table, Valentines cookies, St. Patty's Day clovers, etc. It's a fun way to bring seasons and holidays appropriately into a secondary context.

Tip #2: Answer The Questions

Once students have created questions, pick a few to actually solve together. This will help students feel validated in your class and increase their feelings of belonging in mathematics.

Tip #3: Make It A Competition

Have students work in teams and ask them to come up with as many questions that begin with "How many" as possible. The team with the most questions wins.

Estimation Tasks

This structure is also a wonderful way to bring the real world into our classrooms and give math meaning outside of the classroom. The key idea with this structure is to get students to practice their estimation skills, something that is a surprising challenge for many. You can bring a physical object in that students can make estimates about, for example, number of pretzels in a bag, number of toilet paper squares on the roll, and ask students to create an estimate using the prompt, "About how many…" For example, "about how many pretzels are in the bag" or "about how many toilet paper squares are on the roll?" A wonderful resource is www.estimation180.com with many estimates done for you. The authors of the website encourage you to ask students not for just one estimate, but for an estimate that is too low, an estimate that is too high, and an estimate that is just right. This structure is not just about making an estimate, it's also about asking students to explain why they think they have the best estimate. This is where mathematical thinking comes in and it will blow you away. The point is not to get the exact correct answer (in many cases that's almost impossible) the point is to think critically and mathematically about the context., The anxiety in participating is low because it's not about one correct answer, it's about reasoning and getting an estimate you feel confident about.

Example prompts:

★ Grocery store: Take pictures of interesting items at the grocery store and use them to estimate (cheese puffs, chips in a bag, etc.)
★ Hardware store: Take pictures of chains and have students estimate how many are wrapped around the coil
★ Your house: Take pictures of interesting object around your house for estimation (number of lights on a string of Christmas lights, number of diapers in the box, length of time to burn a candle, etc.)
★ Song length: Show an image of a song player (like iTunes) with a portion of a song already played. Be sure to cover how much time is left in the song. Ask students to estimate the total length of the song

Activity structure:

◆ Show an image (ideas above) to students and ask them the given estimation question (for example: About how long is the song? About what percentage of the pie has been eaten? etc.)
◆ Ask students to share their estimation of the answer AND to explain why they think that is the answer
◆ Scribe as many estimation guesses and reasoning examples as possible (build excitement and interest)
◆ Ask, "want to know the answer?" They will scream, YES!
◆ Do a video or photo reveal if possible
◆ Celebrate estimate guesses that were close, but more importantly the reasoning that led to their estimates

The following are Estimation Tasks tips.

Tip #1: Make It Seasonal

There are lots of ways to bring in seasons and holidays with this structure in an age appropriate way for older students. You can ask for estimates of sprinkles in a jar, how long it will take all of the candles on the menorah to burn out, how many candy hearts in a bag of valentine's day candy, etc.

Tip #2: Competition

Play a "Price is Right" style game where students who get the closest to the answer without going over win a prize. Watch engagement soar.

Table 4.1: Six Core Engagement Structures Resource Table

Structure	Key Idea	Additional Resources
Which One Doesn't Belong?	Has an entry point for all students and is an activity with multiple correct answers, making it a safe structure to participate in.	www.wodb.ca
Notice And Wonder	A game-changer for ensuring students understand the context and information within a word problem or task before they're given a question to solve.	Can be used with any word problem or task. Just remove the question.
Same And Different	A great way to think about a concept multiple times and in multiple ways. With its focus on reasoning, it is safe and approachable for all students to participate in.	www. samebutdifferentmath. com Book: *Same But Different Math*, Sue Looney
Would You Rather	A fun way to encourage decision making and justify reasoning.	www. wouldyourathermath. com
How Many	A fun way to get students to see everyday images in mathematical ways.	Book: *How Many?*, Christopher Danielson
Estimation Tasks	A great way to get students to practice their estimation skills which are so needed for math in everyday life.	www.estimation180.com

You can access done for you resources for each of these activity structures in the chapter resources section at www.gatebreakerbook.com.

Complete Game Changer

Before learning this framework, the struggle was real. We scored as the third lowest school in the state of Wisconsin on the FORWARD exam. I struggled with homework or any kind of practice completion and there was little to no participation in the classroom. I did direct instruction for a good part of the class and then worksheet practice time was the norm for me. Students would just sit and talk during practice work, nothing was getting done.

I registered for the digital PD programs Juliana offers because I'm on a quest to be the best teacher I can be in order to inspire some level of appreciation for math throughout a student's life, even if they don't "love" it. Implementing the strategies from the framework was a complete game changer.

Now, students are excited to come to math class, they are willing to speak and engage in the work of the day, and they complete the tasks on their own instead of using PhotoMath or copying from their peers. These methods have gotten kids talking about math and sharing real life examples of something they saw or talked about with their families over the weekend. My students feel empowered and capable of helping their peers with math and many now enjoy attacking a math problem, even if it is something they haven't learned before. I often hear them saying what I model while I'm thinking, "Where can we start with this problem? What do we know?" Overall I have noticed an increase in accountability, honesty, and work ethic with my students as well as improved test scores.

I especially enjoyed learning about and implementing the routines to boost confidence from the framework. I like that each student, no matter where they are at in their math journey, can have an access point to the content. While these situations can often be unnerving for more introverted students, I have seen tremendous growth in their ability to speak and communicate mathematically. Because our classroom culture is safe, students are willing to volunteer and also encourage their peers to share what they know.

Last year my students had an older teacher who has since retired. He had about a 30% participation rate in his class and growth on district and state assessments was stagnant. Now, since implementing these methods, my class participation rate is 85% and 65% of my students improved their scores on district assessments. I only wish more staff would be trained to utilize these methods and strategies.

Jessica, Title 1 Middle School Math Teacher in Wisconsin

☑ Gatebreaker Homework

Complete the four activities from this chapter with your students:

- ◆ Which One Doesn't Belong without math then with math
- ◆ Notice And Wonder without math then with math
- ◆ Same And Different without math then with math
- ◆ Would You Rather without math then with math
- ◆ How Many
- ◆ Estimation Tasks

You're on a roll! Once you've completed these tasks with students you are done with Phase 1 of the B.R.E.A.K it™ Math Intervention framework: student engagement. You should be feeling the positive community of your classroom building and now you're seeing student engagement and mathematical confidence rising with the Six Core Engagement Structures. You are ready to move into Phase 2: student achievement. This is going to be good!

Reference List

Boaler, J., & LaMar, T. (2019). Valuing difference and growth: A Youcubed perspective on special education. *YouCubed*. https://www.youcubed.org/wp-content/uploads/2019/02/SPED-paper-3.2019-Final.pdf.

Danielson, C. (2019). *How many?* Routledge.

Looney, S. (2022). *Same But Different Math*. Routledge.

Matthews, L., Jones, M., & Parker, Y. (2022). *Engaging in Culturally Relevant Math Tasks 6–12: Fostering Hope in the Middle and High School Classroom*. Corwin.

PHASE 2: STUDENT ACHIEVEMENT

Phase 2 Step 3
Engage Every Student

Phase 2 Step 4
Advance Your Expectations

Step 3: Engage Every Student

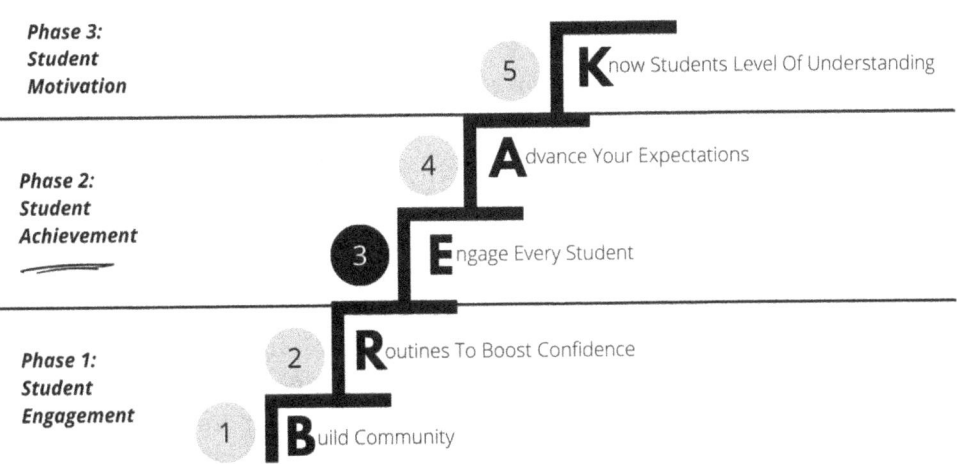

Figure 5.1 Phase 2, Step 3 Of The B.R.E.A.K. it™ Math Intervention Framework: Engage Every Student

I Need To Hear More People Participating

I recently found a thumb drive that had a bunch of files from my time in graduate school at UCLA. One of the files was a video of my first year of teaching filmed by my grad school supervisor. When I first opened it

DOI: 10.4324/9781003479703-8

and began watching, I laughed for ten minutes straight. The shushing, the yelling over the hum of off task talking, the students with their heads down on the desk, the constant posing of questions in hopes of a choral response. Finding this eleven-minute clip from 2011 brought me back to just how challenging that first year was, but also reminded me how much I have grown as a teacher. The first seven minutes of the video contain my direct instruction lesson on graphing in slope intercept form and the remaining four minutes contain footage of an activity where students walk around with graphs taped to their back and have to ask each other yes or no questions to guess the equation of their line. In those seven minutes of direct instruction, I asked 25 questions. One student, Deondre, shouted the answer twelve times. Twelve times! Another student, Brian, shouted the answer three times. I heard a girl's voice shout the answer only twice. I said "shhhhh" seven times and called out specific student names to stop talking five times. Then the comment that really hit home, "I need to hear more people participating."

This video paints a very accurate picture of my entire first year of teaching. Me attempting to lecture, my students not really paying attention, no one really learning. Despite the clear challenges apparent from this short video clip during my first year, it was also a year of tremendous experimentation in my classroom. I had an unwavering belief in my students' ability to do math at high levels and an amazing support system at my school and in my grad school cohort. I kept trying and tweaking things to find what would work for my students. I don't remember exactly what book I had read or what conversation I had with a colleague that spurred the idea, but I realized that with just a small tweak to what I was doing, my students would have more accountability, ownership, and collaboration. I tried it and sure enough, it worked! Over the next few years, and two more high schools in East San Jose and Denver, I would tweak and tighten this instructional method that I named The Math Wars Method® – only because I was a big *Star Wars* fan and wanted to show a clip of the movie in class – until the gains my students were achieving were almost unbelievable. Gains not only in benchmark scores or end of year assessments, but gains in their mathematical confidence and perseverance.

"She is expertly skilled at creating a learning environment where all students feel valued and are motivated to learn because their teacher believes they can," one of my instructional coaches wrote on my evaluation after my third year in the classroom. I would never have believed such glowing remarks from an administrator were possible considering my classroom started as such a trainwreck. It was possible for me and it's possible for you too. I've now trained hundreds of teachers from all over the U.S. to use the

Math Wars Method® effectively in their 6–12th-grade math classroom and the results have proven effective in rural schools, urban schools, and suburban schools. They've proven effective with teachers I support in alternative education schools, facility schools, and schools within the department of corrections. On average, over 500 Math Wars Method® teachers have seen 20% more students passing their classes within just one semester and 46% more daily student engagement. If you want to be a gatebreaker, this is the way to fast track your success. Let's learn all about it.

Engage Every Student

Now we're ready to begin in Phase 2, Student Achievement, at Step 3, Engage Every Student. In this chapter I'm sharing my powerful instructional model, the Math Wars Method®, so you can deliver your lessons in a way that makes sense for your students who struggle, learn powerful strategies to truly engage every student, and no longer feel like you're talking to yourself all period long. I know how painful it is to deliver your lesson only to realize no one was paying attention or understood what you said so you spend the whole period going student to student to student answering the same questions over and over and over again, not to mention struggling with all of the off-task behavior that ensues when students are lost. This method will put an end to those frustrations for good.

The Big Picture Of Math Instruction

There are, generally speaking, two frames we use for thinking about how to teach math: Inquiry-based and traditional or explicit teaching.

Traditional Or Explicit Instruction

In this approach, teachers lecture to teach the information then have students work on practice problems independently. Gradual release of responsibility (I do, we do, you do) is often considered a traditional instructional approach, as is explicit instruction where teachers model step by step problem solving, asking students to repeat. In John Hattie's meta-analysis of 252 influences and effect sizes related to student achievement direct instruction and explicit teaching strategies are rated as having an effect size of 0.6 and 0.57 respectively, both of which are above his identified "hinge point" of 0.4 (Hattie, 2017) suggesting that these methods are effective strategies to increase student achievement. The main challenge with this style of teaching is that while it sometimes shows effectiveness in the short term, long term retention is often missing (Illustrative Mathematics, n.d.).

Inquiry Based Instruction

Inquiry based instruction refers to an approach to instruction that is student-centered and seeks to encourage deeper conceptual understanding instead of the memorization of steps and algorithms. In this approach, students are actively involved in the learning process by deeply engaging in real-life problem solving tasks (Riegle-Crumb et al., 2019). Inquiry based instruction is being integrated in many popular math curricula like Illustrative Mathematics (www.illustrativemathematics.org) and popular PD books like Building Thinking Classrooms (Liljedahl, 2020), among others. Not only does an inquiry approach focus on a conceptual understanding of math concepts, but studies have shown that students are more interested in learning this way, have more self-efficacy regarding their math ability, and see the connection to math in the world around them (Riegle-Crumb et al., 2019). Challenges to inquiry based instruction include student buy in, classroom management, and creation of appropriate tasks (Ernst, Hodge, & Yoshinobu, 2017).

Math Intervention Specific Considerations

Having a classroom full of students who have failed math for many years and are stuck in the gatekeeping cycles of mathematics has unique challenges, and because of this some specific considerations need to be made when choosing an instructional approach with our math intervention students.

Classroom Management

I have been in a lot of 6–12th grade math classrooms as a district instructional coach and independent consultant. One thing I've noticed in classrooms that are successfully using the inquiry based method is strong classroom management. For inquiry based instruction to be effective students must discuss with each other, come back as a whole group for clarification at times, respect differing ideas, and practice other communication skills. This is particularly challenging in math intervention classes where classroom management is almost always a struggle. Additionally, math intervention classes are almost always given to brand new math teachers who typically struggle with classroom management even if it's not an intervention class. Classroom management is challenging in traditional classrooms as well, mainly because the model doesn't create much engagement and when engagement is low, misbehaviors are high.

Math Teacher Shortage

Math intervention classes are typically not very desired teaching assignments and because of this they are often given to the newest hires. Currently most states are experiencing a teacher shortage and even more states are specifically experiencing a math teacher shortage. New math teacher hires sometimes do not possess a math teaching license or certificate. Many of these teachers are being assigned to teach the math intervention course with little to no training in math pedagogy. Throw in a brand new looking inquiry based textbook and the success with an inquiry based approach is going to be highly unlikely. If teachers do not have any math pedagogy training it makes a lot of sense that they would just teach math the way they learned math (that's what I did too even though I had a master's degree in education!) and that is most likely a more traditional lecture plus worksheet approach, which then kicks off the classroom management issues discussed above.

Training

I have seen many schools adopt a textbook with an inquiry based approach, but due to high PD costs and low numbers of teachers using the book (most high schools only have two to four math teachers) they don't get proper training for their staff on how to teach an inquiry based approach effectively. Teachers are left feeling helpless and frustrated with a curriculum that feels foreign to them. This creates a lack of buy-in and teachers often turn to the internet for supplemental or replacement material that ends up being more traditional anyway.

Multilingual Learners

Math Intervention classes typically have high numbers of multilingual learners. Inquiry based instruction is incredibly text rich and can be overwhelming and time consuming for students who need to translate their assignments word for word. Traditional instructional methods are usually more straight to the point, making it more accessible for multilingual learners.

Confidence At The Start

I have seen this time and time again in my own math intervention classroom: The struggle to get started. Teaching math to students who have failed math for multiple years has unique challenges that not even other "regular" math teachers can fully understand. Getting students to jump into problem solving in an inquiry model without explicitly modeling a problem can be almost paralyzing and send students into fight or flight mode, kicking off the disruptive behaviors and classroom management chain reaction.

Which One Should You Choose?

I believe how you teach should be your choice whether it's inquiry based or traditional or, more realistically, some combination of the two, but I have one caveat: It must be effective for your students who are struggling. If you love to teach an inquiry based model and your students who struggle are making gains, keep doing you! If you love a traditional approach and your struggling students are achieving at higher levels, way to go! However, if you picked up this book because your current instructional methods aren't working for your students who need you the most, you need to change your approach.

As educators we talk a lot about equity vs equality. We push for equity, giving students what they need to be successful, instead of equality, giving every student the same resources, and the math instruction debate should be no different. Math teachers shouldn't be forced to teach all students or classes the exact same way, that's math equality. Math teachers should be able to choose the educational experience their students need in order to be successful, that's math equity. If your math intervention students need a more explicit classroom experience because that is what works for them, then provide it. If it's problem based or inquiry based that they need for success, then provide it. If your students need a mix of inquiry based and explicit teaching, then make it happen.

> *Math teachers shouldn't be forced to teach all students or classes the exact same way, that's math equality. Math teachers should be able to choose the educational experience their students need in order to be successful, that's math equity*

I'm ready to ruffle a few feathers. I don't believe the inquiry based approach is most effective for students who are multiple years behind in math. At least not all the time. And this voice, the realistic math intervention teacher voice, is missing from the math instructional debate stage. While it is absolutely essential to integrate tasks and real world examples into your math class for a variety of reasons including cultural relevance, application to the real world, and so many more, I believe a full time inquiry based approach in math intervention isn't what will catch students up to grade level the fastest. And catching students up to grade level quickly is my priority as a gate-breaker. This is not an instructional strategy for all secondary math teachers, this is an instructional strategy designed to be the most effective for students who have been historically unsuccessful with mathematics.

However, I also don't believe the traditional twenty minutes of "I do" aka lecture, followed by twenty minutes of "we do" aka worksheet practice, then twenty minutes of "you do" aka homework is effective for students who struggle either. It *is* boring and does little to include students as capable doers of mathematics. As I shared in my humbling first year classroom video at the opening of this chapter, I have tried this style of teaching with catastrophically horrible results and seen the disengagement and disruptive behaviors it almost always creates in classrooms I've observed in.

A New Suggestion

The Math Wars Method® swings more to the traditional side and closely models a gradual release of responsibility approach, however it presents a small tweak in the traditional system that changes everything and brings in some of the profound benefits of the inquiry approach. It empowers students to guide the progression of problem solving, it creates space for peer to peer collaboration, and it positions the students as the holders of knowledge. Again, The Math Wars Method® is not intended to be the way all 6–12th grade teachers teach mathematics, it's intended to support math teachers in making their lessons accessible, approachable, and empowering for students who have been struggling with math for many years.

Let's first look at the research behind the strategies in this instructional approach before we decide if it's the right fit for you and your students: Cold calling, true formative assessment, gradual release of responsibility.

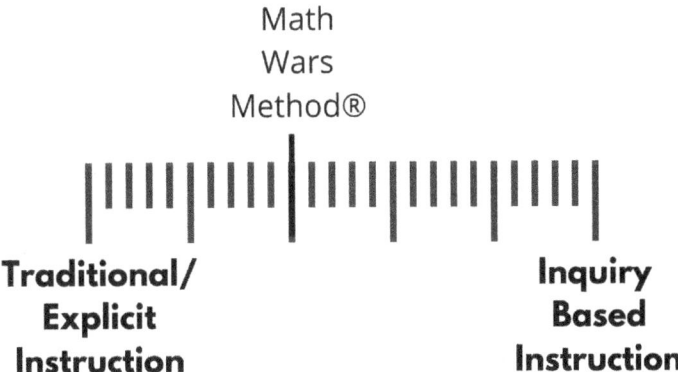

Figure 5.2 The Math Wars Method® Style Of Instruction Leans Closer To Traditional Instruction, But With Some Benefits Of Inquiry Based Instruction

Cold Calling

I'll be honest, I first learned about cold calling when my principal forced us to use it. He had every teacher in the building read Doug Lemov's (2010) *Teach Like A Champion*, and implement Cold Calling as Lemov lays it out. In recent years the book has gained some negative attention for its Eurocentric practices, however I believe the cold call strategy still prevails as the number one way to increase participation in our math classrooms. Cold calling, as defined by Lemov, is to call on students whether or not they've raised their hands. This sentiment is echoed by researcher Dylan William who says:

> when teachers allow students to choose whether to participate or not – for example by allowing them to raise their hands to show they have an answer – they are actually making the achievement gap worse, because those who are participating are getting smarter, while those avoiding engagement are forgoing the opportunities to increase their ability.
>
> (William, 2011, p. 81)

The purpose of cold call is not to "catch" students off task or disengaging, it is to effectively check for understanding from a random sampling of students. This provides you with helpful information about student understanding in the moment, giving you an opportunity to respond to any misconceptions and gaps in the moment. Additionally, research by Dallimore, Hertenstein, and Platt (2012) found that implementing a cold calling system actually increases voluntary student engagement, but maybe more interestingly, students' comfort participating in class in general is higher in classrooms that utilize the cold calling strategy.

I can feel your eyes rolling, your frustrated sigh, you throwing this book to the floor in disagreement, but here me out. Our students who struggle with math are never going to wake up one day and just start participating. We must make participation the norm in our classrooms and, as is the title of this chapter, engage every student if we want to see student achievement in mathematics rise. The best way to do this is with cold call. Many math teachers immediately dismiss cold call because of the trauma it caused them or the trauma they fear it will cause in their students. However, if you've created the community from Chapter 3 and have boosted student confidence by using the Six Core Engagement Structures from Chapter 4, you will not trigger math anxiety or math trauma by cold calling on students. **If you haven't implemented the strategies in** Chapters 3 **and** 4, **do not start cold calling.**

It's very likely you'll do more harm than good if you haven't built a safe community first.

Our students who struggle with math are never going wake up one day and just start participating

---

The video I found of my first year of teaching paints a picture I see very commonly in math classrooms around the U.S. Teachers posing questions to the whole class, no doubt hoping for a choral response of engaged students, only to be met with one or two students actually participating. We think that if we just keep asking the questions the other students will eventually jump in, right? The problem is that it never happens. Cold call eliminates this problem and creates a setting where all students are thinking about the problem because they know they might be called on to answer.

To get started with cold calling, pass out index cards to all of your students. Have them write their name in the middle and decorate it with drawings or words about them. Doing the cards this way makes this a community building and get to know you activity to help you know your students more. Collect the cards and use them all year long to cold call on students.

The following are some common cold calling mistakes.

Figure 5.3 Cold Calling Cards Created By Students As A Community Building Activity Then Collected And Used For Cold Calling All Year Long

Photography by author.

Not Building A Safe Community

I cannot stress this enough: If you haven't taken time to build community as I laid out in Chapter 3 and begun using the routines to boost confidence in Chapter 4, do not jump in with cold call or the instructional strategy I'm presenting here in Chapter 5. Only (if I could publish an entire page of the word "only" here, I would) use cold call if students feel safe to make mathematical mistakes in front of you and their peers. If there is laughing or snickering when students share an answer or especially when they share an incorrect answer, your classroom is not ready for cold call. If teachers do not create the safe community students need before implementing cold call, you run the risk of triggering past math trauma, igniting math anxiety, or worse, creating math trauma for students.

Not Using It Consistently

While it might feel contrary to what you think will happen, research has found that the more you use cold call, the more voluntary participation you will receive (Dallimore, 2012). So I suggest going all in with cold call if you teach students who have been historically unsuccessful with mathematics. Every question. Every day. Lots of teachers ask open questions, expecting a choral response, but in reality it's either completely silent, full of awkward wait time, or the same three students shouting answers. Other times teachers ask questions and expect students to raise their hands to answer, but we all know in a classroom full of students who struggle you'll get the same three or four hands for every question while everyone else zones out, leaving you to feel like you're talking to yourself all period long. Use cold call for every question, every day and you'll never battle these participation woes again.

Not Breaking Down Questions

Do not use cold call to have one student walk through an entire problem. Instead, break the problem into as many small questions as you can and cold call on different students for every one of them. When I see teachers cold call on one student to walk through an entire question I can feel the student's anxiety skyrocket. When I see a teacher call on one student for the first step, another for the next, and so on, I feel the class settle into a cadence and comfort.

Accepting "I Don't Know"

If you allow one student to say "I don't know" and just accept it and move on, you will never see success with the cold calling system. So what should we do when we get that response (because you absolutely will)? First, attempt to scaffold back the question to something the student can answer. For example, if you're asking for the first step of solving 2x - 5 = -6 and you get "IDK" you could ask, "I want to move the -5, what's the opposite of -5?" If that doesn't work, keep scaffolding back as much as you can so that the student can experience success. If they absolutely will not give you an answer, move on to the next student in your cold call deck to answer, but immediately come back to the original student and ask them to repeat the answer so that you are still technically getting an answer from them. This is another *Teach Like A Champion* strategy called "No Opt Out" (Lemov, 2010).

Believing You're Good At Randomly Calling On Students Without Cold Call

Many teachers think they are good at randomly calling on students without using cards or popsicle sticks with student names so they don't need to use a cold call system. The issue with this is two-fold. On one hand implicit bias may be at play and you may be subconsciously calling on specific students more often. Maybe you're calling on boys more often, maybe you're calling on white students more without knowing it, maybe you're subconsciously choosing students who you think will get the question correct. Using a randomized cold calling system is the only way around this. Secondly, you have a million decisions to make as a teacher each day, deciding who to call on doesn't need to be on the decision list!

 Implicit Bias Connection Point

When teachers rely on volunteers to answer questions or think they are calling on students randomly without a physical system like popsicle sticks or index cards with student names, they are more likely to subconsciously fall prey to their own implicit biases and call on male students, white or Asian students, or "smart" students. Using a physical cold call system eliminates implicit bias from sneaking in because every student has an equal, and random, chance of getting called on for every question.

True Formative Assessment

Dylan William's book, *Embedded Formative Assessment* (2011), is a true gem and a book I recommend often. The term "formative assessment," is one of those education-ese phrases that commonly gets thrown around, but often incorrectly. William defines formative assessment as, "Encompassing all those activities undertaken by teachers, and/or by their students, which provide information to be used as feedback to modify the teaching and learning activities in which they are engaged" (William, 2011, p. 37). We typically miss that last part, "to be used as feedback to modify the teaching and learning." Formative assessment isn't this big formal thing that needs to be written down, graded, and recorded, although it can be. Formative assessment at its best, true formative assessment as I like to say, is informal. The key is that we must change our lesson plan based on the formative feedback we gather in the moment. We need to collect informal formative feedback during our class period and then modify our teaching when we see our students aren't getting it. This is how we close the achievement gap. This is how we catch students up in math. This is how we make math equitable. This is how we become gatebreakers. We gather data and we act on it immediately.

This is how we close the achievement gap. This is how we catch students up in math. This is how we make math equitable. This is how we become gatebreakers. We gather data and we act on it immediately.

To modify our teaching in real time sounds time consuming and confusing. First, it's not. One of the recommendations for effective formative assessment in William's book is actually cold calling and I couldn't agree more. If you randomly call on a student and they get the answer wrong or are totally lost, there's a very good chance that they are not the only one who is confused. Instead of feeling disheartened in that situation, be happy! You just gathered wonderful formative assessment data that your students don't get it and now you have the opportunity to respond right then, in the moment, not weeks later when they fail a quiz or test. If you can stick with that student and scaffold back the question to better understand where the confusion is starting from, you are modifying your teaching in the moment to meet students where they are and help them bridge to where they need to be. If you rely on volunteers who already know the answers, your data is skewed because you're not pacing the lesson with all students' understanding in mind and you're missing the opportunity to help identify and bridge the gaps in real time. Secondly, utilizing true formative assessment is vital if we are serious about breaking the gatekeeping cycles of mathematics. William shares data about the effectiveness of many formative assessment studies in his book, but here's his summary, "Students with which the teachers used formative assessment techniques made almost twice as much progress over the year"

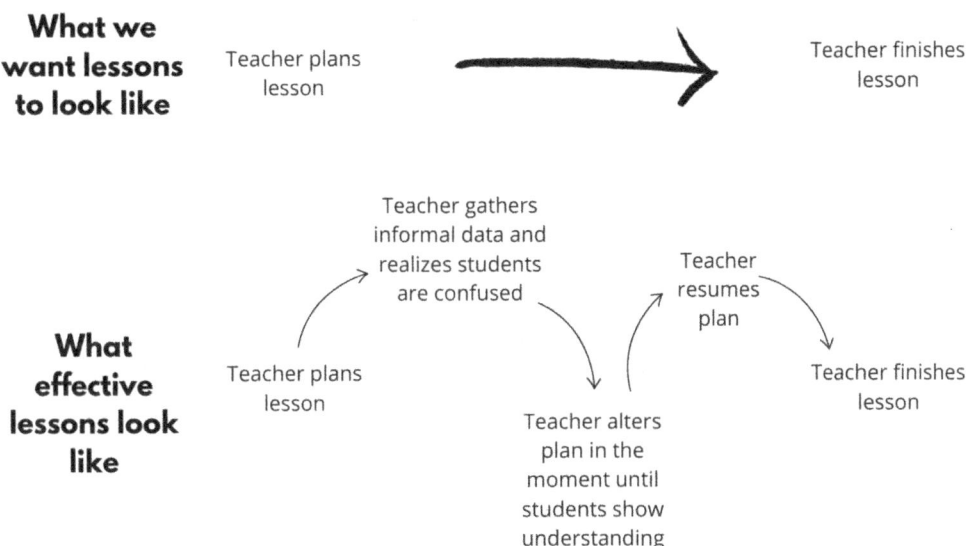

Figure 5.4 What We Want Lessons To Look Like Vs What Effective Lessons Actually Look Like When We Utilize True Formative Assessment Techniques

(William, 2011 p. 37). If we want to be gatebreakers, we must utilize the power of formative assessment in our daily teaching practice and help our students make huge gains in their progress.

Gradual Release Of Responsibility

Commonly referred to as "I do, we do, you do" the gradual release of responsibility (GRR) model is an instructional model where teachers deliver lessons in intentionally structured chunks. In their book, *Better Learning Through Structured Teaching* (2013), Authors Douglas Fisher and Nancy Frey define gradual release of responsibility as an, "instructional framework [that] is based on the belief that teachers can intentionally increase students' ownership of learning over time" (p. 2). They define each part of the framework as:

Focused instruction → "I do it"
Guided instruction → "We do it"
Collaborative learning → "We do it together"
Independent learning → "You do it alone"

Much of the research on GRR, including Fisher and Frey's work, focuses on English Language Arts. Fuentes and Casinillo (2024) have found it to be an effective instructional strategy in science, however research about its effectiveness in math is less established. Some studies have pointed to GRR's immediate effectiveness in test scores, but that the model did nothing to impact retention of the skills learned (Saligumba & Tan, 2018). Yet others show more promising results for GRR as an effective intervention instructional approach (Casas et al., 2023).

The Problem With Gradual Release Of Responsibility (GRR) In Mathematics

The truth is that GRR gets a bad reputation in the math world these days. I get it. Twenty minutes of "I do," followed by twenty minutes of "we do," followed by twenty minutes of "you do," is not effective for students who struggle with math. I know this because I tried it and it did not go well. For one, twenty minutes of direct instruction is too long for many students to stay focused. Additionally, there is no opportunity for students to get immediate feedback about their understanding or make meaning with peers in that time. Of course this isn't working. Of course our students are bored out of their minds. Of course our classrooms are in chaos with little to no work going on. We're not making the content accessible to our students. Many math leaders and academic researchers are moving away from explicit and direct instruction models like gradual release of responsibility in favor of an inquiry based approach and I believe this is because most math teachers misuse the gradual release of responsibility model in their classrooms, negating its effectiveness. One strength of using GRR in mathematics for students who struggle with mathematical confidence is that it offers a predictable structure. Structure and routine is a trauma-informed best practice and if you have students who have experienced math trauma, it's important that students have a predictable structure in your class. The security of seeing a question modeled then gradually releasing the responsibility on to the students to do the math helps students build confidence in their math abilities. I've seen it time and time again, students who walk in with every expectation of failing math class yet again begin building confidence and trusting themselves as mathematicians little by little until all of a sudden, it's making sense and they've built a new relationship with mathematics. Effective use of GRR can pave the way for this transformation.

 Math Trauma Connection Point

Trauma informed research consistently discusses the importance of predictable structure and routine as a best practice. If we make our classroom routine structured and routine as laid out in the Math Wars Method®, you are helping students to overcome their prior math trauma and build a more positive and confident relationship with mathematics.

Table 5.1: Research Base Comparison Chart For The Math Wars Method®

Research	Pros	Cons
Cold Calling William (2011) Dallimore, Hertenstein, and Platt (2012)	*Makes participation the norm* Students who struggle are unlikely to participate on their own, making participation mandatory and the norm will create a culture of 100% participation which is what our students need. *Wonderful informal formative data* If we cold call on a student and they are confused or get the question incorrect, that's great formative data that many students are likely confused. If we cold call on a student and they get it right immediately, that's good formative data that students are understanding.	*Potential trigger* Cold calling absolutely has the potential to trigger math anxiety and math trauma. Before implementing cold calling teachers must build the community laid out in Chapter 4.
True Formative Assessment William (2011)	*Catches learning gaps in real time* When we use informal formative assessment data to modify our teaching in the moment, we catch gaps in real time, not after a test or quiz.	*Can feel overwhelming* The idea of adjusting our lessons on the fly as we take in and process informal formative assessment data feels like a lot to do in an already challenging class.
Gradual Release Of Responsibility Saligumba and Tan (2018) Casas et al. (2023)	*Time efficient* This method of delivering content is extremely time efficient and time is one thing we need as teachers of students who are multiple grade levels behind.	*Often used ineffectively* 20 minutes of "I do" (lecture) then 20 minutes of "we do" (worksheets) then 20 minutes of "you do" (homework) is ineffective for students who struggle.

While it may feel overwhelming to integrate all of these methods together – cold call, formative assessment, and gradual release of responsibility – my Math Wars Method® takes them all and combines them into a three-part framework that can be duplicated in any classroom, for any math content, with any curriculum. I've trained hundreds of 6–12th grade math teachers to use this method and on average teachers see 20% more students passing their class within one semester of using this instructional method. Without further ado, The Math Wars Method®.

The Math Wars Method®: The Prerequisites

Before implementing the Math Wars Method® you must have completed the homework in the previous chapters. This method includes cold calling, sharing your work with peers, and collaborating with classmates which can all be extremely triggering for students who have struggled with math for years. **You cannot skip to this chapter and anticipate any kind of success. You must first build the safe community that the Math Wars Method® needs to succeed.** If you haven't read and implemented the strategies from the prior two chapters, stop right now and go back.

When you're ready to implement the Math Wars Method® you'll need to first make sure your classroom is physically set up for a successful implementation of the method. The Math Wars Method® has an element of collaboration and competition so groups of three to five students are best. If you have some really challenging classes from a management perspective (and believe me I've been there), partners or rows work too. Just make sure students have two to four people they can check their answers with during the period. I like to have mixed ability groups with at least one student who usually "gets it" quickly so they can help the students who typically struggle a bit more. Some teachers worry about cheating and copying in mixed ability settings, but in my experience, students get so into the Math Wars Method® that it really isn't a problem.

To set your classroom up Math Wars style, you're going to need:

1 *Team numbers labeled*: Hang signs above the teams because stickers on desks are likely to be peeled off.
2 *Seat numbers labeled*: Students need to know if they are seat number 1, 2, 3, or 4 (or more or less). I suggest a poster in the front of the room that shows students their seat number since stickers get scraped off.

3   *Team number cards:* One card for every team number you have.
    You can write these on index cards to keep it simple.

4   *Seat number cards:* One card for every seat number you have.
    For example if each group has 4 seats you need cards with the
    numbers 1–4 on them. Again you can make these out of index
    cards).

5   *An area on the board to tally points for each team during the period.*

6   *Individual student name cards or cold call cards:* Have students write
    their name largely on an index card, then decorate it with informa-
    tion about them; their favorite color, favorite sport, something they
    did over the summer, etc. Collect these cards and use them to cold
    call on students.

7   *Optional*: Token economy system tickets or printed money (for ex-
    ample mine was Cummings Cash before I was married and Tapper
    Treasure after).

Once you've got this all set up in your classroom, you're ready to start planning for each lesson using the Math Wars Method® with the steps outlined in
the next section.

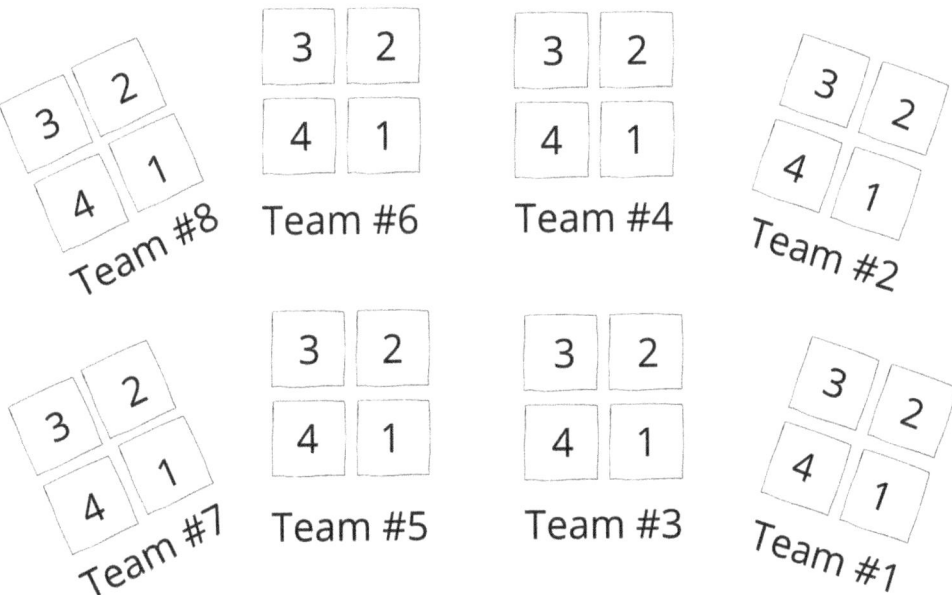

Figure 5.5 Math Wars Classroom Set Up With Team Numbers And Seat Numbers Identified For Students

Figure 5.6 Math Wars Pocket Organizer With Randomized Seat Number And Group/Team Numbers
Photo courtesy of Robin Nehila.

The Math Wars Method®: The System

Step 1: Plan Content

The Math Wars Method® can be used with any content or curriculum. You may need to pull in additional examples, as many textbooks lack adequate examples of what students who struggle need, but hundreds of teachers across the U.S. have used the method with their existing curriculum. If you don't have a curriculum and backwards plan from an exam or benchmark test, you can also use the Math Wars Method® to plan your lessons using the

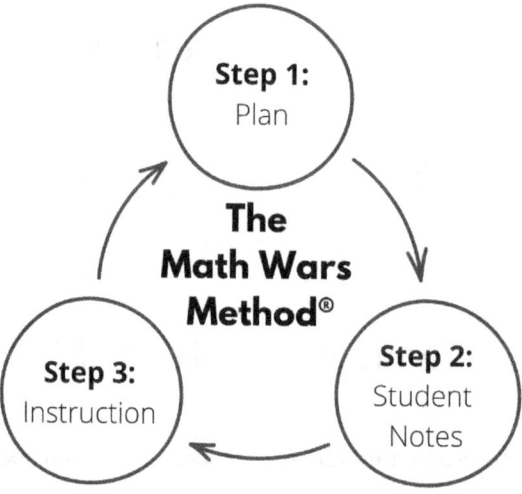

Figure 5.7 The Math Wars Method® Three-Step Approach To Lesson Design

steps below. Lastly if you have no curriculum and no benchmark exams you can still use this method to plan the content for your classes. To lesson plan using Math Wars Method® you need to follow the steps below.

Identify One To Three Chunks Of Content To Teach In The Lesson

A chunk refers to problems that are extremely similar in rigor and solving style. For example solving one-step equations with addition and subtraction would be a chunk (some students might even need addition and subtraction to be separated into two chunks), then solving one-step equations with multiplication would be another chunk and solving one-step equations with division would be yet another chunk. Just because they are all "one-step equations" doesn't mean it's all one chunk. Our students who have struggled with math for years often need a concept broken down this finely for it to finally click for them.

Find Or Create A Total Of Four Practice Problems For Each Chunk

Look in your textbook for additional practice problems or search online. The problem with many textbooks is that they don't break down or chunk the material enough for students who struggle so this is where we have to add our own teacher touch and find what is needed for our students. This part is really important: The four problems need to be EXTREMELY similar in difficulty. If the first problem is $2x = 10$ and the third question is $x/7 = -14$ that is not going to help our students who struggle. Those are two very different problems and need to be in two different chunks.

These four problems will become the "I do, We do, You do, You do" problems during our lesson following the gradual release of responsibility model that is integrated into the Math Wars Method®. By the end of this step you should have the following:

Problem 1 chunk 1
Problem 2 chunk 1
Problem 3 chunk 1
Problem 4 chunk 1

Maybe you have enough time to get through more problems during your class period. If so, you'll have:

Problem 1 chunk 2
Problem 2 chunk 2
Problem 3 chunk 2
Problem 4 chunk 2

And so on for chunks 3 and 4 if possible.

Teachers always ask me how many chunks to plan for and the answer is, it depends. If you're teaching one-step equations you should get through quite a few chunks in a period because those problems are relatively quick. If you're teaching solving systems of equations by substitution you might only get through two chunks in a class period because those problems take a long time to solve. Once you begin using the method, you'll get a better feel for how much content you can cover in a period.

One very important thing to remember here is to always value and accept multiple ways of problem solving. While repetition of similar problems is important in this method, what's *more* important is to value different ways students may solve the problems. If students solve the problems differently than you model, it's okay, it's actually more than okay, it's great! Just because we're going to follow a gradual release model doesn't mean students can't think independently.

Another thing I encourage teachers to do with the Math Wars Method® is make an effort to bring in conceptual understanding of topics whenever possible before teaching the content in this structured way. Again, if I'm teaching one-step equations, the first thing I would do is try to make the concept as concrete as possible with a physical scale and have a more informal exploration of how to figure out how much one of the objects on the scale weighs. Something to pique students' interest in the topic and help the visual learners make the connection between the concrete and the abstract. Once you've allowed some time for exploration, you can move into the structured Math Wars Method® to help them see solving modeled, make their own meaning, work with peers, and grow their confidence.

Once you have your chunks of content planned, let's learn how to have our students set up their notes for success with the Math Wars Method®. ·

Step 2: Student Notes

Grab a student edition of your math textbook off the shelf and turn to page 47. Any student edition of any textbook to page 47. Imagine you're a student who has failed math for years and years and years. How do you feel looking at page 47? My guess is that there is an overwhelming amount of math on that page or an overwhelming amount of words, probably some red ink, probably some arrows pointing to various equations, likely some fractions, am I right? Students who struggle feel overwhelmed whenever they open a textbook or preprinted workbook. They feel like giving up, like pushing the book away and saying "I ain't doing this, Miss," like doing what they've always done in math class: absolutely nothing. In Chapter 1 we talked about the reality of math anxiety and when we ask our students to open to page 47, or 82, or 143 to take notes or complete questions 1–20 odd, that math anxiety skyrockets through the roof!

So how do we use the resources and textbooks we have without overwhelming our students who struggle? Start with a blank sheet of paper. You can use copy paper, lined paper, or graph paper if you're in a graphing unit. Have students fold the paper into eight boxes (I tell them hot dog way once and hamburger way twice) like the image shown below. It's a clean slate every day and it will not overwhelm your students who struggle, plus it doesn't include a trip to the copy machine for you!

Sometimes you will want to preprint a graphic organizer for students. If your particular lesson has a lot of definitions that day you'll want to preprint those for students, especially if you have any multilingual learners in your class, students with dyslexia, or a processing difference that makes it challenging to listen to words and write words at the same time. If your lesson has diagrams you'll want to preprint your notes so that students aren't spending too much time copying the figures, but make sure to white out some of the information on the figures so students can't work ahead of you. The idea here is that students are all working on the same problem together and cannot move on to the next question until everyone moves on to the next question together. If you're preprinting a graphic organizer for notes with your students, make sure some of the information is held back so that everyone can work on the same question together.

Figure 5.8 Math Wars Method® Style Of Notetaking With Paper Folded Into Eight Boxes

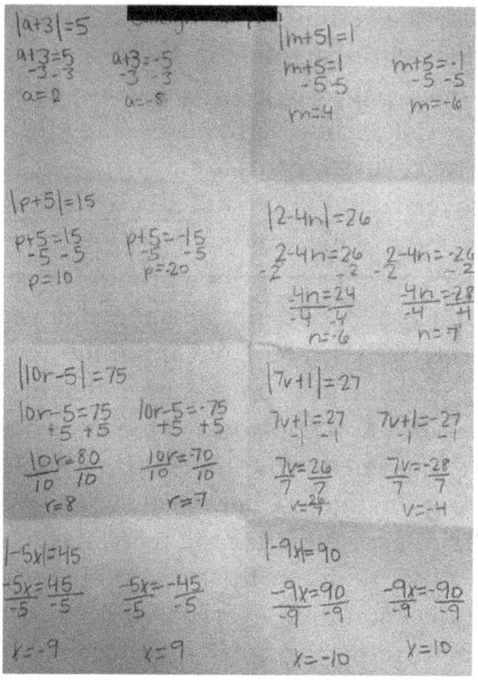

Figure 5.9 Example Of Students Completed Notes

Photo courtesy of Sarah Strange.

⟳ Math Anxiety Connection Point

Research has shown that when students with math anxiety see numbers, problem solving and working memory centers in their brain are compromised. Preprinted notes, student workbooks, and textbooks are completely overwhelming for students with math anxiety due to the amount of numbers, math, arrows, red text, and other annotations covering the page. To help our students overcome their math anxiety we must make notes approachable, not overwhelming.

Step 3: Instruction

Here's where it all comes together. This is the Math Wars Method®. We're going to take the content we planned in Step 1 and deliver it in a way that is accessible for all students while they take notes on the paper we created in Step 2.

Problem 1 Chunk 1: "I do"

Take Problem 1 Chunk 1 in Step 1 and display just that one problem to your students. You can put the folded paper from Step 3 under the doc cam and write along with them, write it up on the board, or get fancy and make a slide for the problem. Remember to show just ONE problem, this is key. You'll solve this one problem start to finish, thinking out loud for every single step and writing notes as you go. Count on your fingers, glance over at the multiplication table, talk to yourself, you're modeling how to problem solve and students need to be able to see it and hear it. Instruct your students to take their folded paper and in the first box, write the notes along with you.

Problem 2 Chunk 1: "We do"

Next, take Problem 2 from Chunk 1 in Step 1 and display just that one problem to your students. This time grab your cold call cards and ask as many questions as possible only using the cold call cards to get answers. Instruct your students to take notes in the next box on their folded paper. For example, if the question is 2x - 5 = -8 here are the questions you might as to a different student for each one:

◆ Student 1: Connie, what do I do first?
◆ Student 2: Great, add 5 to both sides. Juan, what do I get when I add 5 to both sides?
◆ Student 3: 2x = -3, awesome. Mirian, what do I do next?
◆ Student 4: Divide both sides by 2, yes. Dwight, what does that give me?
◆ Student 5: -3/2, Sofia do you agree with that?

In that one question I've heard from five random students, my lesson went at a quick pace, and I have wonderful informal formative data about what my students are understanding and not understanding in the moment, not weeks later when I give a quiz. Of course when you start you'll get some reluctance, you'll get a lot of "IDK", you'll get wrong answers all throughout the year too. This is all okay and to be expected. To respond to reluctance or "I don't know," go back and read the Cold Call section just before this section in this same chapter.

Problem 3 Chunk 1: "You do together"

Now take Problem 3 from Chunk 1 in Step 1 and display just that one problem to your students and ask them to show their work in the next box on their folded paper. Here's where the magic happens. Set a timer – that students can see – for an appropriate amount of time that will create urgency, but not stress students out and heighten math anxiety. Tell students to do this problem together with their team, check their answers with each other, and ask each other for help because, "you don't know who I'm going to call on to answer this one." Circulate and look at student work while the timer is counting down. What do you notice? Is everyone getting the same answer? Are students all over the place with their answers? That is great formative feedback for you about whether or not your students are understanding the material in the moment! When the timer goes off grab those seat number and team number cards and choose a seat number first, then a team number. Ask for that student's final answer to the question and write in on the board. Yes, I said *final answer*. So often, I see teachers set a timer for students to work independently, but then just go over the problem from start to finish when the timer goes off. What reason does this give our students to do any work independently? Why would they do any heavy mental lifting when they know you're just going to do it for them in two minutes anyway? Instead, create authentic accountability by asking for a final answer and fighting the urge to walk through it again step by step. Once you have a final answer on the board, use the cold call cards to call on another student and simply ask, "Do you agree or disagree?" Put all of the ownership on your students. You do not say "yes," "no," "good job," or "not quite." Put the ownership on your students to agree with their peers or disagree with their peers, allow them to be the holder of the math answers, not just you.

⌒; Math Trauma Connection Point

Many students have experienced math trauma when being called on for an answer and either getting it wrong or getting laughed at (or both). It's imperative that you give plenty of time for students to discuss and check their work with their team before calling on someone for an answer. We need to have accountability in our classrooms, but we don't want to trigger math trauma or math anxiety by making it about speed.

Problem 4 Chunk 1: "You do together"

Take Problem 4 from chunk 1 in Step 1 and repeat the same process you just did for Problem 3 Chunk 1, another "you do together." I commonly get asked why this one isn't "you do alone" like how the traditional gradual release of responsibility model is laid out and it's honestly because my students love working together and it didn't feel right to me to stop the momentum I had

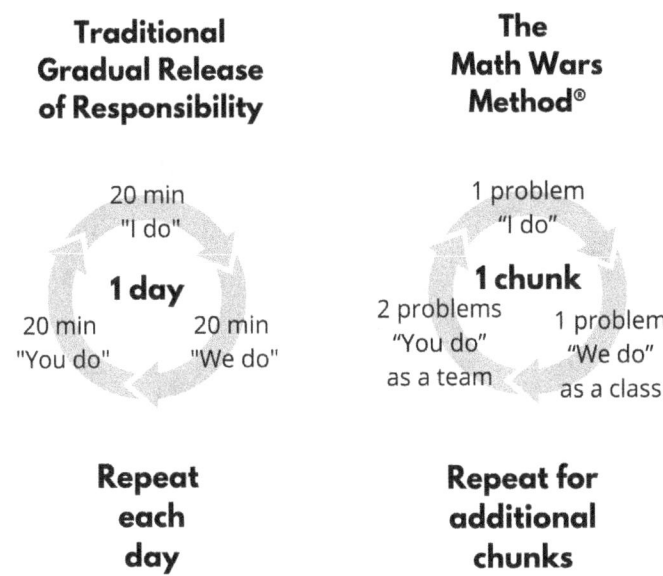

Figure 5.10 Comparing The Difference Between The Flow Of Traditional Gradual Release Of Responsibility With The Math Wars Method®

built in the previous problem by making them work alone. I suggest doing an independent exit ticket at the end of the period or a weekly quiz (more on that in Chapter 7) that is totally independent so that you are able to assess each student's individual understanding.

Then you repeat this process for each chunk:

Problem 1 Chunk 2: "I do"
Problem 2 Chunk 2: "We do"
Problem 3 Chunk 2: "You do together"
Problem 3 Chunk 2: "You do together"

And so on for additional chunks.

That's the process! Maybe you get through three to four chunks in a class period, maybe you get through one or two. Regardless, you use the same process for every chunk.

Learn even more about the Math Wars Method® and get additional support with implementing this system in my online workshop at www.mathwarsmethod.com.

Figure 5.11 Classroom collaboratively working during Math Wars
Photo courtesy of Sarah Strange.

 Math Wars Method® Tips

Suggested Math Wars Method® Points System

I suggest implementing a system to have students earn points for their team to increase the collaboration and add a little competition. For the "you do" problems you can award points to teams who give you a correct answer or confirm a correct answer.

What About A Wrong Answer?

I would be remiss if I didn't stop to address two things you're probably wondering: One, what if the answer is wrong and the other student agrees? Two, what if I get an answer then disagree, disagree, disagree, and I lose track of everything? If the answer is wrong and the other student agrees, give some wait time. Encourage students to call it out if they disagree with the final answer that was given and reward them handsomely for doing so with team points. If we want students to be more independent, confident, and take more ownership, they have to be the ones to trust their work and find the errors, not us. If you know the answer is wrong and no one is saying anything just say, "I disagree," and put the ownership on the students to figure out the error. In regards to the second common concern, if you get a lot of disagrees or see a lot of different answers while you're circulating, it's okay. This is

Table 5.2: Option For Mixing Inquiry Exploration, Explicit Instruction The Math Wars Method®, And Practice Activities For A Typical Week

Monday	Tuesday	Wednesday	Thursday	Friday
Inquiry based lesson to build interest and grapple time	Explicit lesson using the Math Wars Method®	Engaging practice like scavenger hunts, stations, a project, or another inquiry based lesson	Explicit lesson using the Math Wars Method® and engaging practice	Engaging practice and weekly formative assessment (as laid out in Chapter 7)

excellent informal formative feedback that your students are lost. You need to stop and do another "we do" with the cold call cards. You need to find out where students are getting lost and address it right then, in real time, and catch it before you continue.

Frequency Of Using The Math Wars Method®

This is where teacher choice really comes in and knowing what is best for your particular students. I suggest using the Math Wars Method® any time you need to do an explicit lesson. The frequency of using explicit lessons will vary greatly depending on the teacher and the students. Some teachers may only find themselves needing to teach explicitly once a week. Some teachers find themselves needing to teach explicit lessons daily. Use your best judgment and always ask yourself, *is my instructional approach working for my students who struggle the most?* Table 5.2 shows one option for mixing inquiry exploration, explicit instruction the Math Wars Method®, and practice activities for a typical week. However, as I stated earlier you will need to find the balance that works best for you and your students' needs.

Discussion

While this method greatly resembles the traditional gradual release of responsibility model, there are a few key points to make that diminish the disadvantages of the commonly criticized model.

Avoids Stand And Deliver

GRR is commonly criticized for its lengthy direct instructional lecture, during which students are likely to zone out or become disruptive. The Math Wars Method is not that. The "I do" or lecture only takes a few minutes before students are engaged with cold call in the "we do" and "you do as a team" problems. There is one lecture question to every three student guided questions.

Ensures Understanding Before Independent Practice

Another downfall of traditional GRR is that after a lengthy lesson students are then asked to practice independently, but teachers don't have any data about if students have understood the lesson and are ready to practice independently. With the Math Wars Method® teachers spend the "we do" question gathering rich formative assessment data from random students using cold call so that misconceptions have a way to come to the surface and get discussed before students are set to work independently on the "you do as a team" questions.

Elevates Status

In their book *Choosing to See*, which I've referenced many times in this book already, authors Dr. Pamela Seda and Dr. Kyndall Brown discuss the role "status" plays within a math class. Simply put, students who are "smart" have high status and students who struggle have low status. They say, "Inequity is perpetuated by the teacher when the same high-status students are consistently asked the 'harder' questions because they are perceived as the experts" (Seda & Brown, 2021, p. 23). Cold calling on students during the "we do" section elevates every single student's status when done in a safe and supportive math community because everyone has the same chance of being asked harder questions and everyone has the same chance at being perceived as an expert. Every single student will get called on multiple times throughout the period with the Math Wars Method® which helps every student be perceived as an expert and a student capable of doing mathematics at high levels.

Students As Problem Solving Guides

Although the teacher does model one question for each chunk of content in the Math Wars Method®, students are welcomed and encouraged to create their own meaning and make their own mathematical connections during the "we do" and the "you do as a team" questions. Teachers should never mandate that students solve the problems exactly as they did in the "I do" question. Students should always be encouraged and allowed to see math differently than the teacher. If students are confused, they have many opportunities to ask questions and either speed up or slow down the problem solving during both the "we do" and "you do as a team" questions.

Authentic Collaboration

GRR is often considered a Eurocentric approach to learning because it's very individual. Communities of color typically have a more collectivist approach to life and learning. Peer to peer authentic collaboration is an important aspect of culturally responsive pedagogy and is not adequately included in the traditional GRR model. With the Math Wars Method®, authentic collaboration is built into the "you do as a team" problems. Students are asked to work with their team, share their thinking, check their answers, and do error analysis during these problems with the added accountability of just being asked for a final answer that the whole team came to agreement on when time is up. This fosters a community of authentic collaboration, which can be challenging to build in classrooms full of students who struggle.

Students As Knowledge Holders

With traditional GRR, the teacher is positioned as the holder of knowledge imparting it to the students during direct instruction. With the Math Wars Method®, the teacher models and holds the knowledge in only one question. The students hold the knowledge and guide the steps and solving style in the "we do" question and all of the ownership and knowledge holding is put on the students in the "you do as a team" questions because students are asked only for a final answer and another student is asked to agree or disagree with the given answer. This gives all of the power to the students for not only doing the work, but speaking up if they disagree or need clarification. Seda and Brown state, "Inequity is perpetuated by students when they fail to seek the opinions of low-status students in class, even when they have expertise that can contribute to the knowledge of their classmates" (Seda & Brown, 2021, p. 23). With the Math Wars Method® students are given time to collaborate authentically in groups and every student is able to share their expertise with their classmates, elevating any low-status students to high-status students in our classrooms. Because anyone in the group may be called on to represent the group in the "you do as a team" questions, all students get a chance to elevate their status and be seen and heard as capable doers of math.

Use Both Models

As math teachers we can be very black and white, I know I can be, and feel like we have to teach using an inquiry based model 100% of the time or a traditional method 100% of the time. But what if we could be a shade of gray? It's okay to use a little bit of both models. I personally love kicking off a unit with a rich task and letting students take a whole class period to explore the math and get curious. Then I find it helpful for my math intervention students to have some days of explicit instruction to fill gaps (we're diving into this in the next chapter) and get to the grade level content like I show in Table 5.2. You can bring in rich tasks when needed during the unit and again at the end of the unit as a culminating activity or task. It's essential that we're giving students an experience with rich tasks so that they can think deeply about mathematics and develop their critical thinking and problem solving skills, however I also know the realities of teaching math intervention and that there is a lot of ground to cover and explicit instruction is no doubt the fastest way to get there and be a gatebreaker.

Gradually Release GRR During The Year

When students come to us with so much math baggage it can be helpful to start the school year with a more explicit model like the Math Wars Method® and as the year progresses and students grow in their confidence to do math and be mathematical thinkers, begin to integrate more and more inquiry based instruction and less and less explicit instruction, therefore gradually releasing the explicit GRR Math Wars Method® as the year goes on.

This Helped Me Enjoy Teaching More

Before taking the Math Wars Method® workshop, 72% of my students had a D or F. There were numerous students who would not take notes, not even try problems we worked through as a class, not turn in assignments, and I had many students just lay their heads on the desk and/or sleep through classes daily. Only 13% of my students had an A or a B.

I felt like I was pulling a huge weight behind me. No teacher likes to be in a classroom where students dread going. I kept asking myself, "how can I show them that math can actually be fun?" My husband sat through many stories about how difficult it was to get the students to even care about how they did in the class and to take ownership for their performance.

I registered for the Math Wars Method® workshop halfway through the year and I am so glad that I implemented it in my classroom right away. Within just one semester I was amazed at the results.

Second semester I had much better whole class participation and it was a much more positive vibe. Students began to support each other as a team, everyone had an opportunity to share answers with the group, and students even started teaching their peers who needed more help to catch on. Even my administration could see a significant difference.

My D and F rate decreased 25% and now 43% of my students had an A or a B! Plus I loved that most students did not lay their head down on the desk anymore. I am elated that many more students did well in the class. They had a smile on their face when I congratulated them on their grade. I saw that they were putting more effort into completing assignments, which made my job easier. They looked forward to playing the Math Wars game, and asked about it on days we did something different. All this helped me enjoy teaching more.

My life outside of school has been more fun too. I have more time to enjoy with my family and it makes my husband very happy to hear positive comments about school, knowing that I am a happier person.

Marcia, Title 1 High School Math Teacher in Kansas

⊘ Gatebreaker Homework

Implement the Math Wars Method® as outlined in this chapter to make your content readily accessible for all students:

- ◆ Set up the classroom environment (groups, team number labels, seat number labels, cold call cards, etc.)
- ◆ Step 1: Plan content
- ◆ Step 2: Student notes
- ◆ Step 3: Instruction
- ◆ Visit www.mathwarsmethod.com to register for the workshop and get additional support in implementation

Congratulations! Once you begin implementing the Math Wars Method® you are well on your way to increasing student achievement in phase two of the B.R.E.A.K. it™ Math Intervention Framework. You should begin seeing some big improvements in engagement and achievement at this point so make sure you take a moment to celebrate! Now you're ready for the next step in the B.R.E.A.K. it™ Math Intervention Framework: Advance your expectations. This is going to be a game changer chapter so get ready!

Reference List

Casas, A., Casas, M., Evardo, O., & Abina, I. (2023). Integration of Gradual Release of Responsibility Instructional Model (GRRIM) in the Development of Learning Module. *Geometry Journal of Tertiary Education and Learning (JTEL), 1*(2). https://doi.org/10.54536/jtel.v1i2.1778.

Dallimore, E. J., Hertenstein, J. H., & Platt, M. B. (2012). Impact of cold-calling on student voluntary participation. *Journal of Management Education, 37*(3), 305–341. https://doi.org/10.1177/1052562912446067.

Darch, C., Carnine, D., & Gersten, R. (1984). Explicit Instruction in Mathematics Problem Solving. *The Journal of Educational Research, 77*(6), 351–359.

Ernst, D., Hodge, A., & Yoshinobu, S. (2017). What Is Inquiry-Based Learning? *Notices of the American Mathematical Society, 64*(06), 570–574.

Frey, N., & Fisher, D. (2013). *Better learning through structured teaching.* Association for Supervision and Curriculum Development.

Fuentes, A. G., & Casinillo, L. (2024). Assessing the Effect of the Gradual Release of Responsibility (GRR) Model in Teaching Science. *Asian Journal of Assessment in Teaching and Learning, 14*(1), 15–24.

Hattie, J. (2017). Hattie Ranking: 252 Influences And Effect Sizes Related To Student Achievement. https://visible-learning.org/hattie-ranking-influences-effect-sizes-learning-achievement/.

Illustrative Math 6–8 Math Teacher Guide. (n.d). Illustrative Mathematics. https://curriculum.illustrativemathematics.org/MS/index.html.

Lemov, D. (2010). *Teach like a champion: 49 techniques that put students on the path to college.* Jossey-Bass.

Liljedahl, P. (2020). *Building thinking classrooms.* Corwin.

Saligumba, I., & Tan, D. (2018). Gradual Release Of Responsibility Instructional Model: Its Effects On Students' Mathematics Performance And Self-Efficacy. *International Journal Of Scientific & Technology Research, 7*(8), 276–291.

Seda, P., & Brown, K. (2021). *Choosing to see: A framework for equity in the math classroom.* Dave Burgess Consulting, Inc.

William, D. (2011). *Embedded formative assessment.* Solution Tree.

6

Step 4: Advance Your Expectations

B.R.E.A.K. it™ Math Intervention Framework
helping teachers break the gatekeeping cycles of mathematics

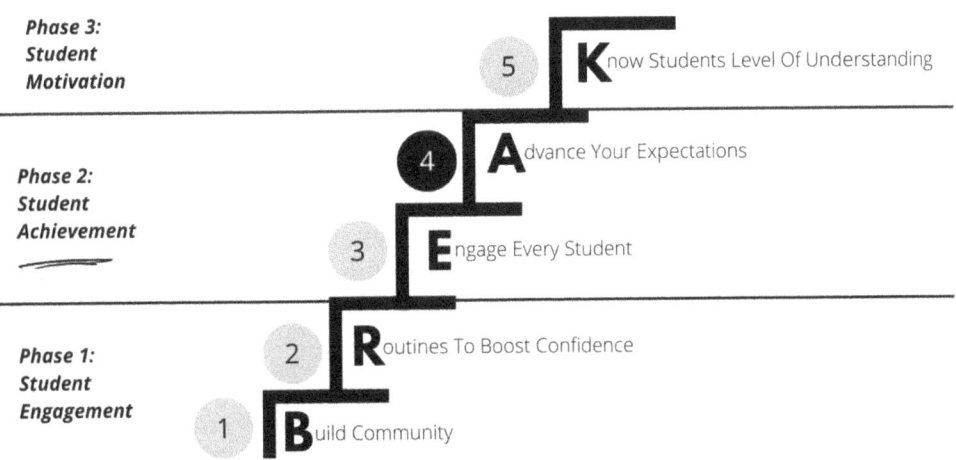

Figure 6.1 Phase 2, Step 4 of the B.R.E.A.K. it™ Math Intervention Framework: Advance Your Expectations

Change Of Plans

"For your 9th grade math support class we need you to just teach the regular integrated Math 1 content instead of the remedial content we asked you to teach," my administrators told me two weeks into the school year starting.

DOI: 10.4324/9781003479703-9

I was teaching high school math support and integrated Math 1 at a high school outside of Denver, Colorado. I was surprised at this request. This was now my fifth year teaching math intervention and my third high school and I was used to teaching remedial middle school content in my math support classes. Afterall, it made sense to do so. We knew that all of the students in the math support class failed one or more years of middle school math and had been socially promoted. We knew they had gaps in their understanding of middle school mathematics and since mathematics builds on itself it makes sense to make sure students understand the middle school content before getting to the Algebra 1 or integrated Math 1 content.

I was upset. Not only had I spent a portion of my summer planning the year thinking that I would be teaching remedial content in this class, but if I'm being honest, I didn't think it was a good idea for the kids. We were already two weeks into the school year and many of them didn't know how to multiply, most of them didn't know how to add negative numbers, and almost nobody knew how to work with fractions, yet my administrators wanted me to just teach them what I was teaching in my regular integrated Math 1 class.

My mind raced with questions:

Could I do this?
Could they do this?
What will I do about all of their missing skills?
How will I teach students how to solve a multi-step equation when they don't know how to multiply?

But since I wanted to keep my new job I said yes. I walked into my math support classes the next day and said, "Starting today we're going to learn Integrated Math 1 in here. I know that you didn't do great in math in middle school so I'm going to make sure that you get the support you need to do the Math 1 work, but I believe you can do Math 1, so I'm going to teach you Math 1." I watched as something changed within each and every student that day. Students who had spent the last few years learning the same thing over and over and over again suddenly heard I wasn't going to print pages and pages of fraction practice, but instead that they'd get the same textbook their "smart" friends had. Something shifted in them from the moment I told them I believed in them. One student, Omar, put it all together and said, "so I'm going to get Integrated Math 1 credit for this class so I can graduate?" When I confirmed that question, his motivation went through the roof! He wanted to pass this class and graduate, but it had felt out of reach being in the support class, now suddenly it felt attainable and he wasn't going to let anything get in his way.

So off we went on our experiment that school year: Could my students who had all failed at least one year of middle school math make multiple years of growth and pass Integrated Math 1 by the end of one school year?

I'll cut right to the chase and give you the results. In general, my math support class performed equally to my "regular classes" and in some cases my math support students outperformed my true integrated math 1 kids on department benchmark exams.

I was amazed and I was forever changed about what math intervention should look like. I don't think I would have ever believed that such results were possible unless I experienced them myself. I certainly would never ask another teacher to run their math intervention class this way unless I had done it myself. But I saw the insane benefits of the power of having high expectations for every single student and now I know I need to share the process with as many teachers as possible so that they can be gatebreakers for as many students as possible.

I share my personal story with you to encourage you. I have truly been where you are, a classroom full of students who have failed math for multiple years, I have tried something that made me uncomfortable and unsure, but it was the best thing I could have ever done for my students. And I encourage you to be open to learning more about this approach to teaching students who struggle with math. It may feel so uncomfortable for you to think about not filling in every skill from elementary and middle school, but will you take a chance with me? Will you keep reading and open your mind just ever so slightly to this different way of thinking about interventions and grade level expectations?

Advance Your Expectations

We continue Phase 2 with Step 4: Advance Your Expectations. In this chapter we'll explore various math intervention models and I'll be sharing exactly how to plan effective interventions whether you're teaching a standalone intervention class or just find yourself needing to provide lots of interventions in your grade level class. An important piece of every math intervention conversation needs to include math fact fluency and basic math skills so we'll be addressing that in this chapter as well. By the end of this chapter you'll have the tools and confidence to say to your students what I said to mine, "I believe you can do grade level math, so I'm going to teach you grade level math."

In 2018 education think tank TNTP (The New Teacher Project) released their "Opportunity Myth" research paper. They sought to understand why students who had met expectations and passed their classes in high school struggled in college. Their research found four key aspects that students need in their daily school experiences in order to be better prepared. I encourage you to check out the full research report, but one key aspect impacted achievement more than any of the others. They found that students who had teachers who did this one thing gained an additional five months of learning compared to their colleagues who didn't do this one thing (TNTP, 2018). So what is this miracle strategy? Teacher expectations of grade level content. That's it. Teacher expectations for students' success against grade-level standards demonstrated the strongest relationship to student growth in the whole study.

The importance of teacher expectations on student achievement has been well documented for a long time. In most teacher education programs we learn about the Pygmalion Effect, a famous study published in 1968 (Rosenthal & Jacobson, 1968). Researchers told teachers that some students were "growth spurters" and thought to be intellectually gifted, but in fact the "spurters" were labeled at random. The study revealed that when teachers expected their students would do well academically, it became a self-fulfilling prophecy and those children did in fact score higher on academic tests than students not labeled as gifted.

This is actually really good news. There is very little we can control as teachers. We can't control what level our students come to us, what their past math trauma may have been, why they have a fear of fractions, if they can do homework each night, the list goes on. One thing we can control are the expectations we have of our students.

The challenge with high expectations for math teachers in particular is that we are bombarded with negative data about our students often before we even meet them. We're assigned to teach the "math support" or "intervention" class full of students who scored "below basic" on assessments the year prior. This makes it all too easy to *expect* little of our students when we're given nothing but deficits to think about. You may not mean to have low expectations of students, in fact I don't believe any teacher sets out to have low expectations of students, but when all the data is so overwhelmingly negative, it can become completely subconscious.

◯ Deficit Thinking Connection Point

When students are grouped into an additional math intervention period or are pulled out of a class for math intervention support, they pick up on their label as "low kid" or "slow kid" in math. These labels – which are often given without conscious thought and have been used casually for decades – are actually a form of deficit thinking because we're labeling the student with their deficits instead of their strengths. Students know when they've been grouped into the "low student" math group and those low expectations often create a self-fulfilling prophecy with low student achievement.

Math Intervention Models

There are two popular math intervention models: just in case intervention programs and just in time intervention programs. Let's dig into each model a bit more.

Just In Case Intervention

With just in case intervention we teach all of the prior grade level content just in case the students have not mastered the concepts. This is by far the most popular of the math intervention models and the easiest to find a "math intervention program" to purchase and use. If you use a math intervention program that gives a diagnostic test then maps out a custom plan to fill in all the gaps a student has from where they score to the grade level they are in, you're using a just in case math intervention model. Similarly, if you teach high school math intervention, but are teaching all middle school content and standards in your class, you're using a just in case math intervention model. This is the model I used at my first two high schools when I taught middle school content in my math support class.

Think about just in case intervention like a street laden with potholes. Each pothole is a standard or concept students missed or failed in a prior math class. The teacher is going around filling every single pothole until a smooth, grade level surface is reached.

Pros Of Just In Case Intervention

Ease Of Use

Just in case intervention models are incredibly popular because of their data driven nature and flashy online software. They can easily be identified with a diagnostic that many publishers and software companies offer with their

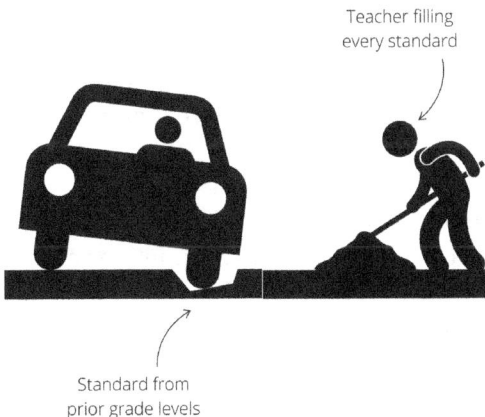

Figure 6.2 The Just In Case Intervention Model

This is an approach where teachers are filling every standard students might have missed in the past, similar to filling potholes in the street.

programs. It's relatively easy to find one of these programs and give math intervention teachers a concrete resource to use with their students. Many administrators are faced with a teacher shortage and particularly a math teacher shortage. These digital just in case intervention programs are easy to hand over to someone stepping in to teach an intervention class even if they don't have a math teaching certification. Teachers are exhausted and overworked, there is no doubt about it and these digital programs are advertised to do it all – the diagnostic, the prep, the planning, the skill progression, which makes it a very desirable option.

Holistic

Just in case interventions are holistic and help ensure students have a strong foundation to build upon when they reach higher levels of math since all of the content they might not have mastered in prior grades is being filled in. Going back to the pothole analogy, we all want a smooth road with no big gaps, the just in case approach will fill all the gaps and make a smooth road of students math knowledge.

Cons Of Just In Case Intervention

Unrealistic Amount Of Material

With just in case intervention programs students usually take the diagnostic, get a grade level score, then the software builds a custom path for them to get from where they scored to their current grade level. The problem is that you're asking students who struggle with math to independently make up multiple years of math that they've already seen and failed time and time again. The sheer magnitude of skills would require a student with tons of motivation, confidence, and ability to persevere and students who are struggling with math are struggling with motivation, math confidence, and the ability to persevere in problem solving. Add in the fact that students are usually working independently on computers for most of these intervention programs and success becomes even more unrealistic.

Math Anxiety Trigger

If your students have math anxiety, looking at a long list of skills they have to master within one year is sure to trigger anxiety and possibly send students brains into fight or flight mode. If you have students who are being disrespectful during computer intervention time this may be the "fight" response to their math anxiety.

Communicates Low Expectations

With just in case interventions you are telling students, "You've struggled with math for a long time and we're going to do all of that math you've

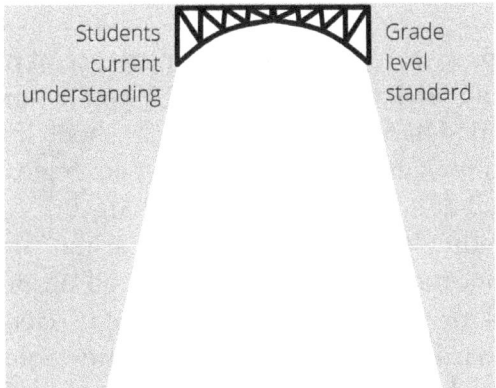

Figure 6.3 The Just In Time Intervention Model
This is an approach where teachers start with the grade level standards and teach interventions and support content needed for success with grade level content, like forming a bridge to grade level content.

already failed again and again and again until you get this time." This is not having high expectations of grade level content and students know it.

Just In Time Intervention

The other type of math intervention model, just in time math intervention, is much less popular. With just in time intervention we teach prior grade level content just before, or just in time for, the grade level content. The name "just in time scaffolding" was coined by professor and researcher Dr. Julie Dixon (2020) and refers to teaching prior grade level content that students might be missing just before, or just in time for, the grade level content. This is the model I used at my third high school when I taught integrated math 1 to all of my math intervention students with additional supports and scaffolds as needed for success with grade level standards.

Think of just in time intervention like a bridge. The goal and destination is grade level proficiency, but students have gaps in some prior grade level skills that are needed to access and master the grade level content. The teacher builds a bridge by teaching content from prior grade levels just in time for mastery of the specific grade level standard or skill.

Pros Of Just In Time Intervention

Communicates High Expectations

With just in time intervention you get to say to your students, "I believe you can do grade level math so I'm going to teach you grade level math." Nothing communicates higher expectations than that.

Expedites Equity

In Chapter 1 we looked at reports on the importance of passing Algebra 1 to open the gates to higher level math classes. We saw how failing Algebra 1 increases the likelihood of dropping out of high school. If we want to increase equitable outcomes for our students, we must strategically create intervention programs aimed at helping students pass Algebra 1. While many of the skills students learn in elementary and middle school are helpful for seeing the beauty and connection of mathematics, not all of them are essential for Algebra 1 success. One example I commonly use is scientific notation. It's a middle school standard but understanding scientific notation has nothing to do with passing Algebra 1. Asking students to spend time mastering standards and skills that don't support the mission of being a gatebreaker and passing Algebra 1 is a waste of time for teachers, but especially for students.

With just in time intervention you get to say to your students, "I believe you can do grade level math so I'm going to teach you grade level math." Nothing communicates higher expectations than that.

Cons Of Just In Time Intervention

Hard To Find

It is much harder to find a just in time math intervention program to purchase because what you teach depends heavily on gathering data from the students you have in front of you at that moment. These programs are less available and less popular.

Takes More Time

Just in time intervention takes more time, there is no doubt about it. With this intervention model we're asking teachers to teach the essential grade level content, but also fill the gaps prior to the grade level content. While a high quality resource might allocate one day for a grade level lesson, a teacher who needs to provide just in time interventions may take three to five days to teach that lesson along with the just in time supports students need for success.

Teacher Mindset Shift

It's hard to ask a math teacher to teach this type of intervention model. There are not many readily available resources so teachers will need to spend time

Table 6.1: Pros And Cons Of Math Intervention Models

	Pros	Cons
Just in Case Intervention	**Ease of use:** Easy to find these intervention programs **Holistic:** No gaps in student skills	**Unrealistic amount of material:** Students are asked to master material independently that they've struggled with for years **Math anxiety trigger:** So much material can trigger math anxiety **Communicates low expectations:** Students are not working on grade level content
Just in Time Intervention	**Communicates high expectations:** Students are working on grade level content **Expedites equity:** Students work on essential content to achieve success in Algebra 1 and beyond	**Hard to find:** Heavily relies on data gathering from your specific classroom **Takes more time:** It takes more time to teach grade level content plus provide the just in time supports **Teacher mindset shift:** Teachers must believe students can do grade level work

creating a plan for their students. On top of the resource creation teachers have to actually believe their students can do grade level work. We all know that students have a keen ability to tell when we're lying to them. If you half-heartedly tell them you believe they can do grade level work, but then your comments and actions don't back it up day after day, the model will fall apart. Using a just in time math intervention approach requires a mindset shift on behalf of the teacher and that is a challenge. This book seeks to help facilitate that shift, but a teacher must be willing to see things differently as well.

Which Model Is Best?

Your "so that" statement from the reflection section in Chapter 1 will likely play a role in which intervention model you feel is best. Since my reason for teaching math is to be a gatebreaker and unlock the Algebra 1 gate for as many students as possible, I firmly believe the just in time approach is the best for our students.

NCTM's landmark book, *Principles to Actions: Ensuring Mathematical Success for All* (NCTM, 2014) illustrates an exemplar scenario of a high school math intervention course to ensure that all students continue to move forward in their mathematics education.

> Multiple design features of the new Algebra Seminar course make it an effective intervention that meets the vision of the Access and Equity Principle. The new course—
>
> (1) teaches Algebra 1 course content along with critical prerequisite content, with a "just-in-time" approach to prerequisite content;
>
> (2) is team-taught by a mathematics teacher and a special education teacher to ensure that the special needs students who are mainstreamed into the class receive the additional support that they need to succeed;
>
> (3) is systematically planned as a back-to-back double period (ninety minutes a day);
>
> (4) is capped at eighteen students, so that teachers have the opportunity to address individual students' needs;
>
> (5) is enriched by focused professional development for the teachers;
>
> (6) uses a broad array of print and non-print, and basal and supplemental, resources;
>
> (7) engages students and enhances instruction with a variety of tools and technology, including interactive whiteboards, graphing calculators, tablet computers, response clickers, and a range of manipulative materials;
>
> (8) incorporates a wide variety of highly effective instructional practices that reflect the Mathematics Teaching Practices; and
>
> (9) draws on online lesson plans and other resources that teachers use to initiate their planning.
>
> (NCTM, 2014, p. 68)

The very first recommendation given by NCTM for an effective math intervention course is the use of a just in time approach. I strongly believe the just in time intervention model is what serves our most vulnerable students the best and will increase equitable outcomes in mathematics so it is the model I will focus on for the rest of this chapter. I know you might be hesitant to try this approach feeling conflicted about the potential gaps in students' mathematical background, but it will be so worth it.

Random fact about me: I'm a huge fan of my Instant Pot. I try to make as many meals in my Instant Pot as possible. It's just so convenient. I can prep dinner, pick up my daughters from school, and come home to a finished hot dinner. A real win in my opinion. Any time I try a new recipe I'm nervous it's going to be awful and I'm just going to need to cook a different dinner from scratch. You might be feeling that way about this intervention approach, nervous that it's going to be awful and you'll need to just start your planning over from scratch. Just like my Instant Pot rarely lets me down, a just in time math intervention approach won't let you down either. If you're willing to try something new for your students who need you the most, let me show you a step by step plan to getting started with the just in time math intervention model.

I realize that sometimes teachers don't have a say in the type of intervention model they can use in their classroom, sometimes the choice is made by administration and cannot be changed. I encourage you to talk with your administrator about the just in time intervention model and urge them to switch to a just in time approach, but in the meantime, read the rest of the chapter and implement what you can, given the constraints of the structure you're being asked to use.

Just In Time Math Intervention Model

Before we dig into the model it is worth noting that this chapter and this intervention model cannot be implemented in isolation and expected to yield results. The prior parts of the B.R.E.A.K. it™ Math Intervention Framework must be in place before implementing a just in time math intervention plan. Teachers must have built a positive classroom community (Chapter 3), be utilizing

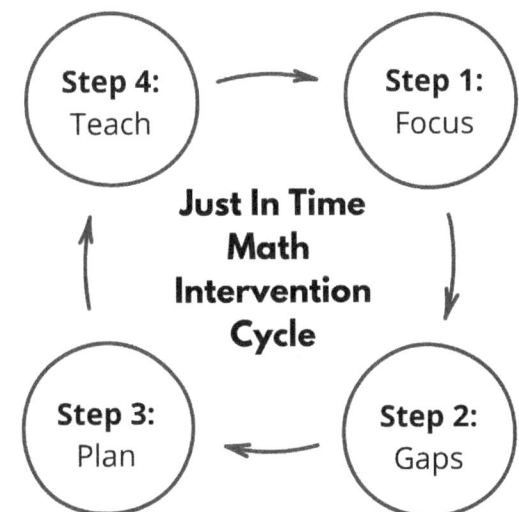

Figure 6.4 The Just In Time Math Intervention Cycle

active participation strategies (Chapter 4), and make the content delivery readily accessible for all students (Chapter 5). Math Intervention success does not begin with *what* you teach, it begins with the *community* you build.

> *Math Intervention success does not begin with what you teach, it begins with the community you build.*

Step 1: Focus

The first step in planning just in time math intervention is to fine tune your grade level focus. You will not be able to get through all of the material and provide the just in time interventions, so instead you must focus on only the content that is essential. You can do this a few different ways. Some schools and districts have already identified key or essential standards for each grade level so check with your administration to see if that document exists. Similarly, some states have already identified key or essential standards for each grade level so check on your state website. Alternatively, you could work with your math department colleagues to brainstorm the essential content together for each grade level. It helps if you make these decisions as a department and come to agreement about what's essential in each grade level as you work to be vertically aligned with each other. Lastly, one of my favorite resources is from Student Achievement Partners (www.achievethecore.org). In 2021 they released their "Priority Standards" for K-8 and a separate document for High School mathematics. Even if this resource is a bit dated, it will still be a wonderful starting point to help you identify the essential content to focus on as well as the content you can let go.

Once you have your essential content identified I encourage you to create a table with two columns like the one below with the first column labeled, "Grade level essential standards." In the first column, make a list of the essential standards or content for your grade level. You can create your own table from scratch using the example below or download done-for-you intervention planning documents for 6th, 7th, 8th, and Algebra 1 at www.gatebreakerbook.com. If you have a textbook or curriculum resource, list out the units and chapters then add the essential standards within each section. This will help you see what to focus on within your curriculum and what to let go. If you don't have a textbook or curriculum resource, still list out the essential content and then find example problems for each of the essential standards. This is a wonderful opportunity to use Artificial Intelligence (AI) to help you come up with grade level examples that illustrate the essential content standard. You can use a prompt like, "give me ten example problems of CCSS. MATH.CONTENT.HSA.CED.A.1" and be amazed as it AI like ChatGPT instantly gives you examples. Additionally you can prompt ChatGPT with "give me five example problems of CCSS.MATH.CONTENT.HSA.CED.A.1 at a DOK 2, a DOK 3, and a DOK 4." As with all AI, do not just assume what it generates is correct or mathematically sound. Use the results as a starting point, but never give students something that was blindly created by AI and assume it is correct or will meet their needs. You must give it your human teacher touch!

 Download done-for-you just in time intervention planning docs for 6th, 7th, 8th, and Algebra 1 at www.gatebreakerbook.com

Step 2: Gaps

Now that you have your list of essential focus content it's time to brainstorm the gaps students might have that must be filled for them to master those essential standards. This is where any diagnostic data would be helpful, but I caution you to not rely on it too heavily as it tends to overwhelm us, sadden us, and has the potential to (subconsciously) lower our expectations of what our students can do.

Taking the table you created in Step 1, label the second column, "Prerequisite Skill Gaps." You can use the table below as a guide or download the done-for-you intervention planning documents for 6th, 7th, 8th, and Algebra 1 at www.gatebreakerbook.com. In the second column brainstorm the gaps you anticipate your students might have as they work up to the grade level content. Let's look at an Algebra 1 example:

> *Multi-step equations are the grade level content, but I know my students likely struggle with combining like terms, distributing, solving one-step equations, and solving two-step equations, all of which are prerequisites for solving multistep equations that are part of prior grade levels standards.*

You will need to go through all of the essential content and ask yourself, what gaps might my students have from prior years that they need in order to access and achieve success at this grade level standard? Each year the prerequisite skill gaps may be a little different. Each period the gaps might be a little different. This is why it's so challenging to have a done-for-you just in time math intervention program. To implement just in time interventions

Table 6.2: Just In Time Math Intervention Planning Table

Step 1: Focus Grade Level Standards	Step 2: Gaps Prerequisite skill gaps
If the concept you're planning to teach is not on this list, SKIP it. You have to make room for all the just in time interventions.	If you're teaching intervention as a separate period, focus on these skills to provide the support just in time for the grade level content. This is not an exhaustive list. You can add more. Ask yourself, *"What skills do my students need to be successful in this grade level content?"*
List essential standards here	*List prerequisite skills here*

effectively, teachers need support, professional development, a library of resources, and most importantly time to create and find lesson content for their specific students.

Step 3: Plan

Now that you've fine-tuned your grade level focus and identified the prior grade level skills that might be missing or weak, it's time to plan how to fit it all in. There are two types of teachers reading this book and they each have a slightly different path in this plan section so let's dig in for both types.

Intervention Teachers

These teachers teach an additional math period or maybe pull students out or push into math classes to provide interventions. These teachers should focus on teaching the list of skills created in Step 2, the prerequisite skills. The plan is to teach the prerequisite skills that will help students be successful in their grade level math class. You don't need to teach everything from prior grades, just the content that is going to help them be successful with grade level content. It's always a plus if intervention teachers can also provide additional practice or an introduction to the grade level content they will see in their grade level math class.

Grade Level Teachers With Many Students Below Grade Level

These teachers are teaching the grade level math class, but have many students below grade level in their classes. These are the 7th grade teachers with a lot of students who are scoring at a 3rd, 4th, or 5th grade level. These are the Algebra 1 teachers with students who haven't passed a math class since elementary school. After the COVID-19 pandemic, these types of classes are on the rise and they are very challenging because of the range of student ability and content level mastery within the class. For this group of teachers, you have a more challenging and time consuming plan. You need to plan to teach the prerequisite skills just in time, or just ahead of, the grade level content. Let's go back to the multi-step equation example:

> *Now that I've identified the prerequisite skill gap content (distributive property, combining like terms, etc.) I must create a plan to teach that gap content right before I teach the grade level content so that I'm providing the intervention just in time for the grade level content. I may plan for a day of combining like terms, a day of distributive property review, a day for solving one-step equations, and maybe two days for solving two-step equations just before my lesson on solving multi-step equations which is the grade level standard.*

You can really see how this takes extra time. Just in time intervention is not a quick fix, it's not easy, but it will get you better results and create a classroom of more engaged, motivated, and confident mathematicians.

The following are Just in Time Intervention Planning tips.

 ## Tip #1: Don't Take Too Long On The Prerequisite Skill Content

As an instructional coach it is not uncommon to walk into an Algebra 1 class in December and see the class still going over basic operations. Students are struggling with fact fluency, there's no doubt about it (in fact we'll be talking more about fluency later in this chapter). However, we do not need to take months to review basic operations. Those skills are part of every single lesson moving forward so students will have plenty of time to master them as you move forward with grade level content. Remember that the biggest benefit of just in time intervention is that it communicates high expectations to students. You can't tell them "I believe you can do grade level math" and then spend months doing math they've seen and failed, seen and failed. You have to move on and trust that they will master these fluency skills with time.

Tip #2: Differentiation

Many of us have classrooms with a mix of students far below grade level along with a good number of students who get it, are bored, and want to move on more quickly. This has always been a challenge in education and it's a multifaceted issue. Since my mission is to be a gatebreaker and to do everything I can to open the gate of Algebra 1 for my students, I will always make decisions based on meeting the needs of my most struggling students first. If you can let students who want the class to be more fast paced work independently with a blended learning model or Modern Classrooms approach (www.modernclassrooms.org), that would be ideal so that you can bring the most valuable resource – you – to the students who need you the most. If a blended learning approach is unrealistic or overwhelming, I believe it's most impactful to set up your classroom with the needs of your students who struggle as the priority.

Tip #3: Diagnostic Data

This intervention approach is perhaps less data driven than you might have expected. I'll be covering assessment techniques in Chapter 7 and a weekly assessment system that can and should be used to inform your just in time intervention planning, however, more helpful than giving a large diagnostic assessment that collects tons of prerequisite skill information is gathering a smaller amount of data at the start of each unit on prerequisite skills needed for grade level success in that particular unit. Teachers can give a short (no more than five questions) assessment to identify which prerequisites students know and which they haven't mastered yet. Keeping the assessment short keeps math anxiety at bay plus it makes grading easier for the teacher. The other reason I find large diagnostic exams to be ineffective for planning math intervention content is mainly because students' prerequisite skills may strengthen as the school year progresses. Giving a massive diagnostic exam at the beginning of the year may have data that is totally outdated by the time you get to the unit that requires those skills. Additionally, I have found that giving a diagnostic exam early on in the school year, before you've had the time to build an intentional community, may give results that don't truly show what the students can do. The amount of effort students are willing to put in on assessments once you've built a trusting and safe community cannot be underscored. By giving small prerequisite assessments at the start of each unit you are giving your math intervention students a chance to show you they are growing and gaining more confidence in mathematics.

Tip #4: A Note For Administrators

You need to allow your teachers to use supplemental material for filling the gaps. While it is essential that teachers have a high quality curriculum for their grade level instruction, in order to be labeled as "high quality" these curricula are actually not allowed to have prior grade level standards within them. While this is wonderful to communicate high grade level expectations, teachers do still need to fill in the content students are missing. In my client work I commonly hear administrators ask teachers to use a new curriculum with fidelity and disapprove of the teachers bringing in additional supplemental materials. It is impossible to provide effective just in time interventions if teachers are not allowed to stray from the curriculum for a day or two to provide targeted interventions just ahead of the grade level content. If you purchased multiple copies of this book for your department, be sure to visit www.teams.gatebreakerbook.com to access additional bulk order bonuses to support your team with implementing a just in time approach.

Step 4: Teach

Finally once we've identified our grade level focus standards, brainstormed the gaps, planned to teach the prior grade level content just in time for the grade level content, it's finally time to teach! I believe the most effective and efficient way to catch students up to grade level is to use the Math Wars Method® that I shared in detail in the prior chapter.

You'll repeat the same four steps each time you plan for your intervention class or each time you need to provide interventions for your students. Implementing a just in time math intervention model requires a huge shift in the mindset of the teacher. Truly believing that all students can do grade level math regardless of how many years of math they've failed is a mental hurdle that is difficult for many to overcome. I don't take it lightly that I'm recommending this challenging approach. I had my own doubts too, but switching to a just in time approach was so immensely beneficial for my students, I could never go back to teaching math intervention any other way. I encourage you to try it, even if it's just for one unit. Give yourself a chance and give your students a chance.

Math Fact Fluency

I would be remiss to write about any kind of math intervention model, but particularly just in time math intervention, and not discuss how math fact fluency fits into intervention. Many teachers have grown increasingly frustrated with the lack of mastery of basic math facts, especially in older grades. I commonly hear teachers at their wits end about how their 9th and 10th graders don't know how to multiply, how their students continue to count on their fingers, and how students are reliant on the calculator for the most basic of operations. I get it. I never had a 9th grade class that knew how to multiply or subtract negative numbers and don't even get me started on fractions, my students froze in fear when a problem had a fraction. While it is frustrating that our middle school and high school students aren't coming to us prepared with a strong mastery of the basics, it's well outside of our sphere of control.

Fluency is defined as, "knowing how a number can be composed and decomposed and using that information to be flexible and efficient with solving problems" (Parish 2014, p 159). Number sense is the foundation that fluency is built upon. Number sense is a student's ability to use and work with numbers in meaningful ways. Students become mathematically fluent when they have a strong number sense. All of our math students, but in particular students who struggle with math, need to develop their mathematical fluency to support their success in grade level math. Afterall, you can't solve multi-step equations if you struggle to add or subtract negative numbers, multiply, or divide. Fluency, and therefore number sense, must be a part of the math intervention conversation.

However, how to build that number sense and develop fluency is a heated debate. On one side people argue memorization and speed of recalling basic facts is the best way to develop number sense. On the other side people argue that focusing on conceptual understanding is the best way to develop number sense. Research on students with learning differences in mathematics established long ago that the traditional drill and kill style of learning number sense is inadequate and at times even counterproductive (Garnett, 1992).

Math Trauma Connection Point

Many students have experienced a traumatic math event related to fluency and number facts. For example, getting a simple math fact wrong publicly, being made fun of for a wrong answer to a basic math problem, or freezing up while taking a timed multiplication test. Students' trauma is triggered any time they see the same type of math they were doing when the original trauma occurred. Once their trauma has been triggered their working memory is diminished, making fact recall nearly impossible, and their primal brain goes into fight or flight. We must be mindful of the ways we ask students to practice their math facts so that we don't trigger math trauma.

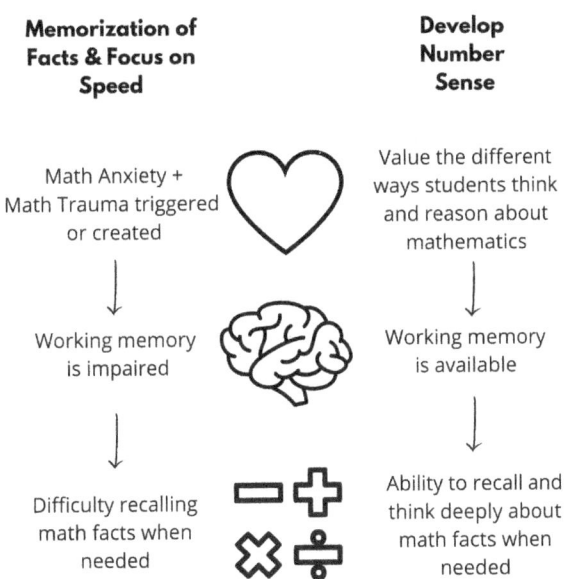

Figure 6.5 Comparing Two Common Ways To Get Students To Build Mathematical Fluency: Memorization Vs Number Sense

Common Misconceptions About Number Sense

Fingers

A lot of math teachers dissuade – or even shame – their students from using fingers to add or subtract. They feel students in late middle school or high school should have their facts memorized and be able to recall the facts quickly without using their fingers. However, there are multiple studies providing proof that the opposite is true. Berteletti and Booth (2015) make a clear point, "Evidence suggests that finger representation and finger-based strategies play an important role in learning and understanding arithmetic" (Berteletti and Booth, 2015, p.109). Let your students use their fingers. You might even want to model counting on your fingers in front of the whole class so they don't feel ashamed if they need to use their fingers.

Speed

There are numerous studies about the importance of disassociating speed from mathematics work. Researchers conducted brain scans revealing that math facts are largely held in the working memory area of the brain (Beilock, 2011). We also know that stress has detrimental effects on working memory (Lukasik et al., 2019). When we put this together it becomes logical that the stress created from a timed test would impede students' working memory and their ability to recall math facts. Students should be able to think deeply about any mathematical question and praise should not be given for fast answers.

Memorization

Many teachers believe that students should have their multiplication facts memorized and quickly be able to answer multiplication questions when asked. Memorization should not be a requirement in our math classes. A study by Supekar et al. had two important findings. First, some students are able to memorize math facts more easily than others. Second, the students who did memorize math facts more easily did not have higher IQ scores, were not any higher achieving, and didn't have superior "math ability" (Supekar et al., 2013). Just because some students can memorize facts quickly and easily does not make them any "smarter" or more capable of doing mathematics.

Building Fluency While Maintaining High Expectations

How we build mathematical fluency is only part of the question. The other part of the fluency question is how do we develop fluency with older students and continue to maintain high grade level expectations? The biggest benefit to a just in time math intervention model is that it communicates high grade level expectations to students. However, practicing multiplication (just one example) in a 9th-grade Algebra class is not a grade level standard. How can the need for fluency practice and grade level expectations exist together? To maintain high expectations and develop mathematical fluency, practice must be time aware, age appropriate, and visually approachable.

Time Aware

You don't need to spend weeks or months ensuring every student has mastery of their basic math facts. These basic skills will be spiraled and integrated into every problem students do in upper grade math. They will have plenty of time to master these concepts as you go through the year and maintain high grade level expectations of your students. Additionally, math intervention classes tend to have a more transient student population. If you take months to review the basics only at the beginning, students who enter your classroom after that point will be at an additional disadvantage. Try integrating number sense routines and activities into your warm ups, creating "fluency Friday" time, or spending one day per unit developing mathematical fluency.

Age Appropriate

It's very easy to print worksheets that were made for lower grade levels and just give them to our older students for fluency practice. For example, if our students need to practice division we might find a division worksheet made for 4th graders that happens to have big bubble letters and butterfly clip art. We pass it out to our 9th graders, but they just crumple it up and throw it in the trash calling it "little kid" and proceed to put their hoodies up and ear buds in and totally check out from class. Can you blame them? They've probably seen that exact butterfly worksheet four or five times by now. They feel disrespected. They certainly don't think you believe in them or their ability to do grade level math. Make sure your fluency practice is age appropriate for your students. Don't worry, I'll be sharing some resources and ideas shortly so keep reading.

Visually Approachable

If you print out a drill and kill basic math fact worksheet, there is almost no way that a student is going to engage with that learning experience. If a student struggles with math anxiety, all of those numbers are literally triggering the fear center in their brain so not only does it look overwhelming, their ability to problem solve and recall facts needed to complete the worksheet is diminished due to the anxiety and fear. That is not a recipe for success. This is not going to help our most struggling students feel empowered to tackle multiplication, fractions, or any other basic skill that has been keeping them behind for years Year after year they see the same problems over and over. They know you are frustrated. They know they are "supposed" to know these things. But at this point it's easier to laugh it off and distract others than open themselves up to learning. There is too much trauma, anxiety, and fear. We can avoid triggering our students' math trauma when we select fluency activities that are visually approachable instead. I'll be sharing my favorite resources soon.

Bonus: Fun

Try your best to make your fluency practice fun! Our struggling students have been asked to practice these skills for years on end, make it different and make it fun in order to increase engagement. I love bringing in a deck or cards or dice to make fluency practice more fun and feel more age appropriate.

Bonus: Conceptual

Park & Brannon (2013) found that when students practiced a non-symbolic (pictorial based) arithmetic task, their ability to compute symbolic arithmetic (like a symbolic equation) improved as well. If we want students to deepen their number sense, we must give them the opportunity to see and think about math visually and conceptually. In the age of artificial intelligence (AI) this statement is more relevant than ever. Students have calculators and AI at their fingertips for quick computational answers so we must focus on conceptual understanding if we want them to think deeply about mathematics – and limit the chances that student responses were created by AI.

The Calculator: To Use It Or Not To Use It? That Is The Question

Part of the fluency conversation must also be calculator use. Do we let our students use a calculator in class or use class time to build number sense without the calculator? The answer is yes. Yes we should build in specific time for students to build number sense without the calculator and yes we should allow students to use the calculator to access grade level content. Here is some more guidance on how to think about the calculator as a tool and when to use it effectively for students who struggle.

Do we let our students use a calculator in class or use class time to build number sense without the calculator? The answer is yes. Yes we should build in specific time for students to build number sense without the calculator and yes we should allow students to use the calculator to access grade level content.

When To Use The Calculator

If a student needs a calculator in order to access grade level mathematics so that teachers can maintain high expectations of them, let them use it. If we are doing challenging grade level math and the student needs a calculator to help with basic calculations in order to make the grade level math accessible, I will always let them use the calculator. The argument against calculators used to be, "well it's not like you walk around with a calculator in your pocket!" But thanks to smartphones, we do. Another argument against using calculators is that students can't use them on state assessments and standardized tests so we're setting our students up for failure on those high stakes tests if we let them use it as a crutch all year long. My rebuttal to this is to teach students to make a multiplication table on their exams. I have found that my students mostly use the calculator to do multiplication and division, so teach them to flip over their test and make a multiplication table before they do anything else. Then teach them to use that multiplication table during the test when they need it. I also encourage teachers to hang multiplication tables up in your room and allow students to look over to the chart when doing multiplication or division every day in class. This will give them the confidence to use a multiplication table during an exam. Lastly, many students with IEPs in math are allowed to use calculators for the vast majority of state assessments so if you're teaching a class with many students on IEPs giving them time to practice with the calculator is actually a huge benefit.

When Not To Use The Calculator

I would encourage you to make fluency practice a regular part of your class routine. When it's intentional fluency time, tell students they can't use the calculator and use activities that focus on conceptual understanding to make the non-calculator time more valuable for your students. For some ideas of intentional non-calculator fluency practice ideas, check out the table below.

 ### Gatebreaker Tools

The following is a list of resources for teachers to find high quality fluency and number sense practice that is appropriate for older students. More recommendations can be found in the resources section of www.gatebreaker-book.com.

Number Talks

What it is: The purpose of a number talk is to get students thinking deeply about arithmetic, be able to explain their reasoning, and appreciate how classmates see mathematics in different ways. For example you could give the problem 18 x 4, give students time to think about the answer (without a calculator), then share their thinking. You'll want to have a few students explain their reasoning and how they saw and solved the problem to validate their mathematical thinking as well as show students that there are multiple ways to think about math problems. While most of us may say we line up the 18 and 4 and multiply the traditional way, there are many other ways to decompose this problem and build fluency with numbers. One student might do 10 x 4 and 8 x 4. Another student might do 18 x 5 then subtract 18. There is no limit to how a classroom full of students will see this problem and that's the beauty of it!

Why it's great for students who struggle: Focusing on thinking deeply about one problem is visually approachable for all students. Additionally, number talks are great ways to increase all students mathematical status in the classroom since all students have a chance to share their reasoning and be seen as an expert. I find using number talks as warm ups once a week or twice a month are a great way to get older students thinking about numbers more conceptually and helping them see that there are many ways to compute these basic facts they've struggled with for so long.

More information: Humphreys & Parker (2015).

Open Middle Tasks

What it is: My favorite Open Middle tasks are the problems that ask students to fill in the blank boxes with the numbers 0 to 9 or -9 to 9. The tasks are like little riddles to solve that encourage deep and conceptual thinking about numbers, patterns, and so much more.

Why it's great for students who struggle: The first thing you should know is that Open Middle Tasks are challenging for students, but also for me and many of the math teachers with whom I share this resource, so expect students to struggle a bit the first time they see one of these problems. Even though the tasks are organized by grade level on the website, don't be afraid to look at earlier grade tasks to use as basic skill practice. Even if you use a prior grade level task, for example double digit multiplication with 10th graders, there is nothing "little kid" about it, it is age appropriate to show to all students. Additionally Open Middle Tasks are visually approachable because students are focused on solving only one problem. Be sure to check out the "Open Middle Worksheet" that author Robert Kaplinsky has available on the website to encourage perseverance with these tasks.

More information: www.openmiddle.com and book: Kaplinsky (2020).

YouCubed Tasks

What it is: A library of low floor, high ceiling tasks searchable by grade level created by YouCubed based out of Stanford University.

Why it's great for students who struggle: Since these tasks are all low floor, high ceiling tasks they have an entry point for all students meaning there is a place for every student to start the task no matter how many days of school they have missed or how far behind they might be. I love the focus on number sense and conceptual understanding. One of my favorites is the "Math Cards" activity to help students make conceptual and pictorial connections with symbolic representation and arithmetic.

More information: www.youcubed.org/tasks.

Desmos Teacher Activities

What it is: Most math teachers are aware of Desmos as a graphing calculator or approved calculator for state tests, but Desmos is so much more! Click over to the "teacher" section and access a free library of done for you activities for any topic and any grade level. These activities use technology to illustrate mathematics and bring it to life.

Why it's great for students who struggle: The way many of these activities can illustrate math is perfect for your visual learners. Many activities have digital manipulatives built in so you don't have to deal with purchasing physical manipulatives making it a wonderful resource for your kinesthetic learners as well.

More information: teacher.desmos.com.

Games

What it is: Make fluency practice fun with games like Bingo or activities that let students use a deck of cards or set of dice.

Why it's great for students who struggle: Instead of pre-printing a page full of math problems only to trigger math anxiety, let students create their own problems using cards or dice. Other traditional games like Bingo (that look appropriate for older students) that practice multiplication, division, or other basic skills are a fun way to spend some class time building number sense and developing mathematical fluency.

More information: I have created many fun fluency games which you can find in the resources section of www.gatebreakerbook.com.

Six Core Engagement Structures

What it is: All of the Six Core Engagement Structures can be found in Chapter 4. They are: Which One Doesn't Belong, Notice And Wonder, Same And Different, Would You Rather, How Many, and Estimation Tasks.

Why it's great for students who struggle: They are all structures that focus on students explaining their reasoning instead of computing one correct answer. They can also all be introduced without math in order to get students familiar with the structure and not trigger math anxiety.

More information: Look back to Chapter 4 for all of the Six Core Engagement Structures or in the resources section of www.gatebreakerbook.com.

80% More Students Passing Algebra 1? Yes, Please!

I was teaching the Algebra 1 lab class made up entirely of students who had failed Algebra 1 the semester before. I felt like I was talking to myself all day. Students would not raise their hands to answer questions and often I would get "IDK" if I asked them a question. On top of that fighting the phone battle was exhausting. I was extremely frustrated and was not happy coming to work on a daily basis.

I found Juliana on Instagram and registered for the Math Wars Method® Workshop shortly after the semester started and my classroom has completely turned around.

Now all of my students consistently take notes and answer questions. We are doing grade level content every single day! My test scores have improved significantly, but the best thing that I have learned is how to engage my students. I have less issues with students disengaging by putting their heads down or playing on their phones.

My principal observed this class for my formal observation and he walked out with me at the end of the period and said, "That was awesome! You had those students eating out of the palm of your hand and they were all engaged!" Now I enjoy coming to work again.

Since I'm teaching the students who failed Algebra 1 last semester, and honestly most haven't passed a math class in years, 0% of my students had a passing grade. Now since using the Math Wars Method® and understanding how I need to set up and teach my intervention class with grade level content, 80% of my students are passing! Within one semester! I owe Juliana a debt of gratitude. This class has turned into my favorite to teach and I'm actually excited to teach next year!

Sarah, High School Math Teacher in Indiana

⊘ Gatebreaker Homework

This was a challenging chapter. Some of the points made might be in opposition to everything you thought to be true about mathematics. If you need to sit with the thoughts shared here before implementing, please do. When you're ready to continue your journey to becoming a gatebreaker here's what you'll need to do:

- ◆ Switch from a just in case intervention approach to the just in time math intervention approach laid out in this chapter
- ◆ Create time for number sense and fluency practice using the activity ideas listed in this chapter

This is a big moment! You are officially done with Phase 2 of the B.R.E.A.K. it™ Math Intervention Framework! You've built the engagement in Phase 1 and student achievement should be on the rise after implementing the material in Phase 2. Only one more phase to go to fully transform your classroom for students who struggle and become the gatebreaker they need you to be! Let's keep the momentum going into the next chapter, the final chapter of the framework.

Reference List

Beilock, S. (2011). *Choke: What the secrets of the brain reveal about getting it right when you have to.* Free Press.

Berteletti, I., & Booth, J. R. (2015). Perceiving fingers in single-digit arithmetic problems. *Frontiers in Psychology*, 6, Article 226. https://doi.org/10.3389/fpsyg.2015.00226.

Dixon, J. K. (2020, November 17). Just-in-time vs. just-in-case scaffolding: How to foster productive perseverance. *Houghton Mifflin Harcourt.* https://www.hmhco.com/blog/just-in-time-vs-just-in-case-scaffolding-how-to-foster-productive-perseverance.

Garnett, K. (1992). Developing fluency with basic number facts: Intervention for students with learning disabilities. *Learning Disabilities Research & Practice*, 7(4), 210–216.

Humphreys, C., & Parker, R. E. (2015). *Making number talks matter: Developing mathematical practices and deepening understanding, grades 4–10.* Stenhouse Publishers.

Kaplinsky, R. (2020). *Open middle math: Problems that unlock student thinking, grades 6–12.* Stenhouse Publishers.

Lukasik, K. M., Waris, O., Soveri, A., Lehtonen, M., & Laine, M. (2019). The Relationship of Anxiety and Stress With Working Memory Performance in a Large Non-depressed Sample. *Frontiers in Psychology*, 10, 4. https://doi.org/10.3389/fpsyg.2019.00004.

National Council of Teachers of Mathematics. (2014). Access and equity in Mathematics education. https://www.nctm.org/uploadedFiles/Standards_and_Positions/Position_Statements/Access_and_Equity.pdf.

Park, J., & Brannon, E. M. (2013). Training the approximate number system improves math proficiency. *Psychological Science*, 24(10), 2013–2019. https://doi.org/10.1177/0956797613482944.

Parrish, S. (2014). *Number talks: Helping children build mental math and computation strategies, grades K-5. Updated with common core connections.* Math Solutions.

Rosenthal, R., & Jacobson, L. (1968). *Pygmalion in the classroom: teacher expectation and pupils' intellectual development.* New York, Holt, Rinehart and Winston.

Student Achievement Partners. (n.d.). Achieve the core (website). https://achievethecore.org/.

Supekar, K., Swigart, A. G., Tenison, C., Jolles, D., Rosenberg-Lee, M., Fuchs, L., & Menon, V. (2013). Neural predictors of individual differences in response to math tutoring in primary-grade school children. *Proceedings of the National Academy of Sciences, 110*(20), 8230–8235. https://www.pnas.org/doi/full/10.1073/pnas.1222154110.

The New Teacher Project. (2018). The opportunity myth: What students can show us about how school is letting them down – And how to fix it. https://tntp.org/publication/the-opportunity-myth.

PHASE 3: STUDENT MOTIVATION

Phase 3 Step 5
Know Students' Level of Understanding

Step 5: Know Students' Level of Understanding

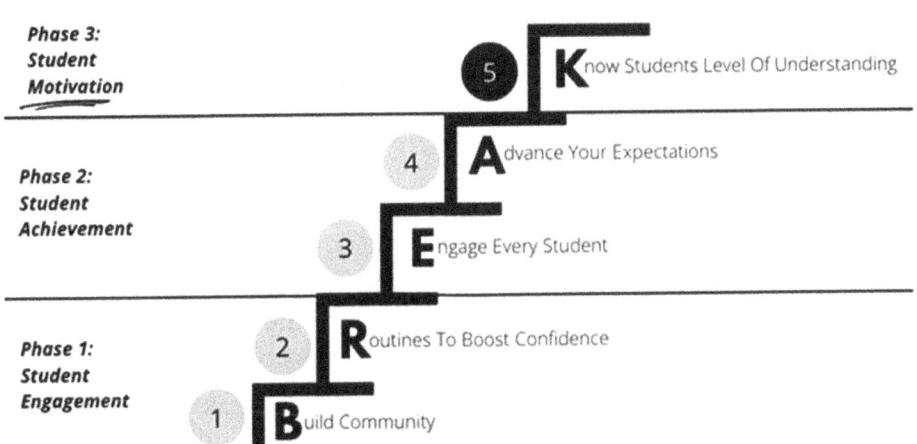

Figure 7.1 Phase 3, Step 5 Of The B.R.E.A.K. it™ Math Intervention Framework: Know Students' Level Of Understanding

They Got Fired

Two of our most experienced teachers got fired mid-year because they wouldn't do this. They wouldn't grade students based on a rubric. At the start of my second year, a new Principal came in and encouraged (cough,

DOI: 10.4324/9781003479703-11

forced) everyone to switch to standards-based grading and grading students with a rubric instead of the traditional 0–100% grading system. In general, we were a very young staff. Eighty percent of us had been teaching less than five years so we were open to trying new things because we didn't have our systems set in stone yet. But two of our most experienced teachers could not get on board with this grading approach. They felt strongly that this type of grading system wasn't what was best for our students so they were fired in the middle of the year.

If I'm being honest, I had my own concerns about this system too. I mean, the 0–100% grading system was the only assessment system I had ever known. Through my entire elementary, middle, high school, college, and grad school experience there was only one way to grade, 0–100%. Questions correct out of questions total. And as a math teacher it makes sense. I wondered if this system was going to create grade inflation. I wondered if students were going to see a 50% grading floor and just stop trying. I wondered if students even cared about how they were graded. But I wanted to keep my job so of course I complied. Afterall, what did I have to lose? It wasn't like the traditional grading system I had used the year before yielded very fruitful results with about 40% of my students failing my Algebra 1 class.

As a staff we read books about standards-based grading and worked together to create a conversion system so that we could give students a rubric score on their physical assignments, but enter it into our gradebook as a percentage (since we still had traditional gradebooks). As departments we collaborated on how this grading approach would look for each of our subject areas. We tweaked, experimented, and ultimately learned so much that first year of implementation.

I was astonished at the results. Every single worry I had was disproven. Grade inflation? My students tried more challenging problems than I ever imagined they would. Fifty percent grading floor? It actually motivated my students more than I ever thought possible when they realized that if they had an off day or one bad test it wouldn't be impossible for them to recover their grade and get back on track. Student buy in? I never knew students felt the 0–100% grading system was unfair and that they would be so open to a new, more equitable and transparent way of grading. In this first year of implementation I was completely sold. I knew I would never go back to any other way of grading. I eventually moved to two more schools and at each school I took this grading system with me and it yielded the same amazing results. Increased student motivation, insane buy-in, and more engagement.

I know you might have some doubts about standards-based grading, 50% grading floors, and grading students with a rubric in math, but I encourage you to keep an open mind to the grading alternative presented in this chapter.

It was the missing piece of the student motivation puzzle for my students and the countless teachers with whom I've now shared this exact same system. Without this missing piece, knowing students level of understanding, everything else you have worked so hard and intentionally to build will crumble.

Know Students' Level of Understanding

We've reached the final phase and step of our framework! Phase 3, Student Motivation, and Step 5, Know Students' Level of Understanding. In this chapter we are talking all about assessment practices and using assessment as a tool to know your students level of understanding. By the end of this chapter you will understand why the 0–100% grading system is inequitable and why a mastery-based grading system is the most effective way to increase motivation for all students, but particularly students who struggle with math. I know how it feels to give a quiz and feel totally heartbroken when you realize almost everyone failed. I know how it feels to give a test and see the class goofing off instead of taking it seriously. I know the pain of seeing students completely shut down when they realize there's almost no way to pass your class. I also know the frustration of having a mound of tests and homework to grade when you're trying to have a life outside of school. But I also know there is a better option. In this chapter you'll learn my three part Rethinking Math Assessment Framework™ so that you can implement a grading system that motivates instead of discourages and helps you be a gatebreaker instead of a gatekeeper. Not to mention the streamlined process will cut your grading time drastically so you can live your life outside of the classroom too. This framework will unlock the final phase of the B.R.E.A.K. it™ Math Intervention Framework, student motivation.

What's Wrong?

To help illustrate what is wrong with the 0–100% grading system and why it hinders student motivation, let's look at three fictitious, but incredibly common student scenarios.

> *Dwight: On test one he scores 3/10. On test two he scores 6/10. Despite doubling his score, he is still failing in a traditional 0-100% grading system. Will Dwight feel encouraged to keep trying to double his score again? Will he keep putting in the effort needed to finally begin passing the class? I find it highly unlikely. I call this the impossible hole.*
>
> *Alejandra: On test one she scores 2/10. On test two she scores 8/10. Despite showing proficiency or even mastery on the second test, her average*

is still a 50% and she is failing the course. I can only imagine how disheart-
ened Alejandra will feel when she realizes that despite proving her knowl-
edge, it's still not enough to pass the class.

Xavier: He has math anxiety and test anxiety and even though he
does well in class, when it comes to tests he freezes as soon as he gets to
the more challenging problems. He gives up and turns in what he has, but
it's not enough to pass. He may stay engaged in class for a little bit, but
after time he will have no reason to give it his all when he can't ever seem
to pass a test.

What's Right?

We need something different, our students need something different. That's
where mastery-based grading, standards-based grading, or rubric-based
grading comes in. I will use all three terms interchangeably throughout this
chapter. Mastery-based grading offers an alternative to the 0–100% system
that does not serve students who are struggling with motivation in mathematics. Instead of thinking of grades as the number correct out of total number of questions on a quiz or test, mastery-based grading focuses on the type
of question (simple, complex, or challenging) students get correct then translates that into a rubric score ranging from zero to four. Let's look at the benefits of mastery-based grading.

Empowerment

Students understand exactly which questions they need to get correct on a
quiz or test in order to pass the assessment. If a student wants to get a C,
they know exactly what they need to focus on and not get overwhelmed by
other questions. Similarly, if a student wants an A or a B, they know what
they need to do to achieve their desired result. This really helps students like
Xavier mentioned above because they are less likely to give up on a challenging question knowing that if they get it wrong, it won't impact their grade as
heavily as with a traditional test.

Equity

Students find standards-based grading to be incredibly equitable because it
is so transparent. Students know exactly which questions they have to get
correct to get an A, B, or C and track their progress each week so they know
exactly what their overall grade should be at all times. There are no surprises.
Additionally, in true standards-based grading the only scores that go into a
grade are assessment scores, no participation grades, no homework grades,
just grades that represent mastery towards standards. With mastery-based

grading there is no way our opinions of a student, our own deficit thinking, or our subconscious implicit bias can impact a student's grade and that is what makes it truly equitable.

Mastery-based grading offers an alternative to the 0–100% system that does not serve students who are struggling with motivation in mathematics. Instead of thinking of grades as the number correct out of total number of questions on a quiz or test, mastery-based grading focused on the type of question (simple, complex, or challenging) students get correct then translates that into a rubric score ranging from zero to four.

Implicit Bias Connection Point

Since implicit bias is subconscious, it's likely to show up while grading student assessments or even more likely when we are entering participation or citizenship grades. If a student misbehaves, teachers often want to take away participation points as a consequence for the disruption. However, if grades are supposed to represent what a student knows about the content, allowing grades to be impacted by something as subjective as behavior is unfair and inequitable. One way to ensure equitable grading practices is to base our students grades only on their math performance with a mastery-based grading approach.

Motivation

A rubric score of zero doesn't mean zero points in the grade book, it actually means 50%. This motivates students and encourages their perseverance because they'll never have to dig themselves out of that impossible hole that Dwight and Alejandra experienced. While there is lots of debate about using a 50% grading floor, when you adopt a mastery-based grading approach it's helpful not to think of student scores by the percentage, but instead by the rubric score. If a student didn't do anything and you feel uneasy about giving a 50% in the gradebook, focus on the fact that they got a zero on the rubric.

Rich In Data

The strategic weekly formative assessment I'll share with you shortly creates valuable and rich data about student understanding. Combining this framework with the informal formative assessment data collected with the Math Wars Method® (Chapter 5) allowed me to eliminate the need to grade nightly homework because I had such a deep understanding of what my students learned during the week that I didn't need homework (which can now be done so easily with AI anyway) on top of it.

Researched-Based

When my principal forced us to change to this rubric-based grading approach we read many different books on standards-based grading to inform the system that we came up with and apply it in our 0–100% gradebooks. The most impactful book was Dr. Robert Marzano's (2009) *Formative Assessment and Standards-Based Grading*. Dr. Marzano is the standards-based grading guru. One big distinction I make in my Rethinking Math Assessment Framework™ is the passing grade. Dr. Marzano describes proficient understanding, and a passing grade, as a level 3 (questioning levels are coming up in the next section). When my colleagues and I adopted this system we made a passing grade a level 2, developing understanding. My rubric is not set in stone. If you need to adapt it or want to use the cutoffs in Dr. Marzano's book, please do.

Rethinking Math Assessment Framework™

 Download the Rethinking Math Assessment Framework™ quick guide in the resources section of www.gatebreakerbook.com

This powerful assessment framework is the most overlooked piece of the student motivation puzzle. I know it was for me and hundreds of other 6–12th grade math teachers who now use this system in their classrooms. There are three parts of the framework that you'll need to understand to implement this system in your classroom. First, we'll explore the question levels. This is the foundation of the entire assessment framework and we'll deeply understand what makes a level two question, a level three question, and a level four question. In Part 2 we will dive deep into the weekly formative assessment, called the "weekly FA." In this part you'll

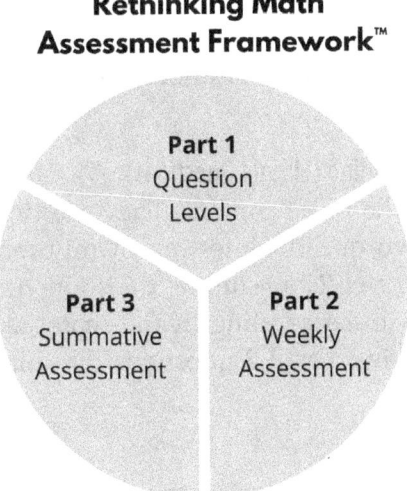

Figure 7.2 The Three-Part Rethinking Math Assessment Framework™

learn how to create the strategic five-question formative assessment or quiz to give each week made up of a formula of specific question levels. You'll also learn how to grade that weekly FA, use the data to create differentiated small groups, and supercharge the framework with student led feedback tracking. Lastly, in Part 3, we'll explore the summative assessment practices you'll need to understand in order to use this framework with any existing summative assessment like chapter tests, unit exams, or district benchmark assessments. One of the most common misconceptions about this framework is thinking that every teacher in your department or school must be using a mastery-based framework, but that is not true. You can use this framework even if you're the only teacher using it and you can still use all of the same assessments your colleagues use. We'll cover aligning the summative assessment with this framework, how to grade a longer summative assessment with this framework, as well as discussing the importance of student reflection and revision. I realize that even *thinking* about changing your assessment practices is deeply personal, but I urge you to keep an open mind as you read this chapter. It might just be the key to student motivation and long lasting student engagement.

Part 1: Question Levels

There are three levels of questions you'll need to understand for this assessment system. They are a part of both the weekly assessments as well as the summative assessments so understanding the different question levels is paramount to the entire framework. There are three question levels: 2, 3, and 4. Each level refers to a specific level of understanding.

Level 2: Developing Understanding (Simple Questions)

Level 2 questions are the basic content. They are typically DOK (Depth of Knowledge) one or two questions testing mainly recall and specific skills. When students get a level 2, or simple, question correct they are demonstrating a developing understanding of the material. To be a gatebreaker, I was comfortable making Level 2 questions the minimum for passing the assessment.

Level 3: Proficient Understanding (Complex Questions)

Level 3 questions are the complex content. They are typically DOK 2 or 3 questions testing more challenging skills than a level two question or involve strategic thinking. When a student gets a level three, or complex, question correct they are demonstrating a proficient understanding of the material. This is what Dr. Mazano requires as the minimum for a passing grade.

Level 4: Advanced Understanding (Challenge or Stretch Questions)

Level 4 questions are the challenge or stretch questions that are related to the content, but take it beyond what was explicitly taught in class. They are DOK 4 questions testing extended thinking. When students get a Level 4, or challenge, question correct they are demonstrating an advanced understanding of the material.

Part 2: Weekly Assessment

The cornerstone of this assessment framework is the weekly assessment. This strategic five-question assessment given every Friday yielded such helpful data about my students' understanding that I felt comfortable letting go of collecting and grading homework. In Part 2 of the framework we'll look at how to create this strategic five-question weekly formative assessment which I called the "weekly FA," how to grade the weekly assessment, how to group students for small group intervention support, and finally how to boost student ownership through feedback tracking.

Table 7.1: Rethinking Math Assessment™ Weekly Assessment Structure

FA Question	Rubric Question Level	Description
#1	2	This question addresses the fundamental skill or concept covered during the week, aligning with Level 2 or developing proficiency.
#2	3	These questions are the meat of what was taught that week. In student-friendly language these would be questions that mean "I know it as well as my teacher taught it." These questions are aligned with Level 3, or proficient.
#3	3	
#4	3	
#5	4	This question is the challenge question where students take the skills and concepts they learned, but apply it to a question that wasn't explicitly covered in class. This question is aligned with Level 4, or advanced proficiency.

Weekly Assessment Structure

Every week you'll follow the exact same format for your weekly FA.

Just as important as the assessment itself is what happens when you give the assessment to your students. Here are some best practices for the weekly FA facilitation.

Week 1 FA: Solving Systems By Substitution

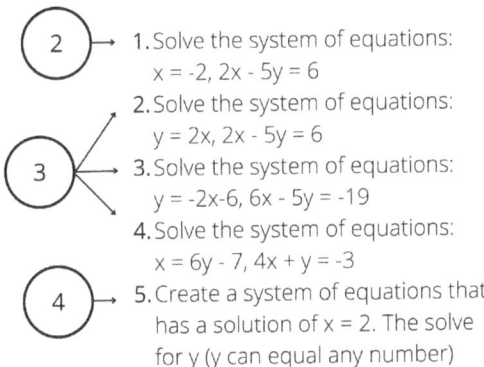

2 → 1. Solve the system of equations:
x = -2, 2x - 5y = 6

3 → 2. Solve the system of equations:
y = 2x, 2x - 5y = 6

3. Solve the system of equations:
y = -2x-6, 6x - 5y = -19

4. Solve the system of equations:
x = 6y - 7, 4x + y = -3

4 → 5. Create a system of equations that has a solution of x = 2. The solve for y (y can equal any number)

Figure 7.3 An Example Of A Weekly Formative Assessment To Show To Students Assessing Knowledge Of Solving Systems By Substitution

Image created by author.

Label The Question Levels

Clearly label each question with its level of difficulty (i.e., Level 2, Level 3, or Level 4). This transparency helps students understand the expectations, motivates them to engage with the assessment, and helps students with test anxiety to feel empowered.

Allow Enough Time

Allow fifteen to twenty minutes so students don't feel rushed. Many students have math anxiety and testing anxiety so you want to make sure that you give them plenty of time to complete the assessment so that they're not feeling rushed. Fridays worked best for me personally, but all school schedules are different.

Prevent Cheating

Since students will likely be sitting in groups (especially if you're utilizing the Math Wars Method® from Chapter 5) it's essential that you actively monitor during the assessment time. Because you are using this assessment to gather rich data to be used to identify and close learning gaps, it's essential that it's a true understanding of what each individual student knows and is able to do. One best practice is to share this with your students, tell them that this is the one thing that's graded each week and it's important that it reflects what you really know. Remind them they can always retake an assessment if they aren't happy with their score. An additional tip is to tell students to use their "off arm" to cover their assessment. An "off arm" is just whatever arm they aren't writing with and it's used to shield their answers from any drifting teammate eyes. Invention of "off arm" credit given to Mrs. Edwards, my across the hall teaching neighbor in South LA, and the greatest science teacher of all time. I would encourage you to say, "I don't want to accuse anyone of cheating, so don't give me a reason to even think that you're cheating. Use your off arm and mind your eyes." Be transparent with your students about the importance of not cheating on this one data point and they will respond well, especially if you've built the community outlined in Chapter 3.

Grading The Weekly Assessment

With mastery-based grading you grade based on the **type** of questions students get correct instead of the traditional grading approach of number of questions correct out of number of questions total. There are five different scores a student can get ranging from zero to four. First let's look at the rubric to percentage conversion table then dive into how to grade the weekly FA specifically.

Table 7.2: Rubric To Percent Rethinking Math Assessment Framework™ Conversion Table

Rubric Score	Gradebook Percent	Description
0	50%	Even with help, the student didn't get anything correct
1	65%	With help, the student had partial success on simple content
2	78%	The student got MOST of the simple questions (Level 2) correct
3	88%	The student got all of the simple questions correct and MOST of the complex questions (Level 3) correct
4	95%	The student got all of the simple questions, complex questions, and MOST of the challenge (Level 4) correct

If you follow the five question strategic assessment formula from the previous step, here's how the grading will break down:

Rubric Score Of 0
The student did not attempt the assessment or got every question incorrect.

Rubric Score Of 1
The student asked for help and needed a lot of scaffolding and support to get any question correct. This is a very uncommon score.

Rubric Score Of 2
Must get #1 correct. Since #1 is the only simple question, if they get it right, they get a score of 2.

Rubric Score Of 3
Must get the majority of #2, #3, and #4 correct. Since #2, #3, and #4 are all complex questions, they must get most of the complex questions correct. If they get only one of those correct, they do not get a score of 3. They must get either two or three of the complex questions correct to receive a score of 3.

Rubric Score Of 4
Must get all questions correct. Since #5 is the challenge question, they must get all of the simple and complex questions correct in addition to the one challenge question.

Data Driven Grouping
Technically, a formative assessment is only a formative assessment if you modify your teaching practices after receiving the formative assessment data (William, 2011). Only when we modify our teaching in real time will we reap

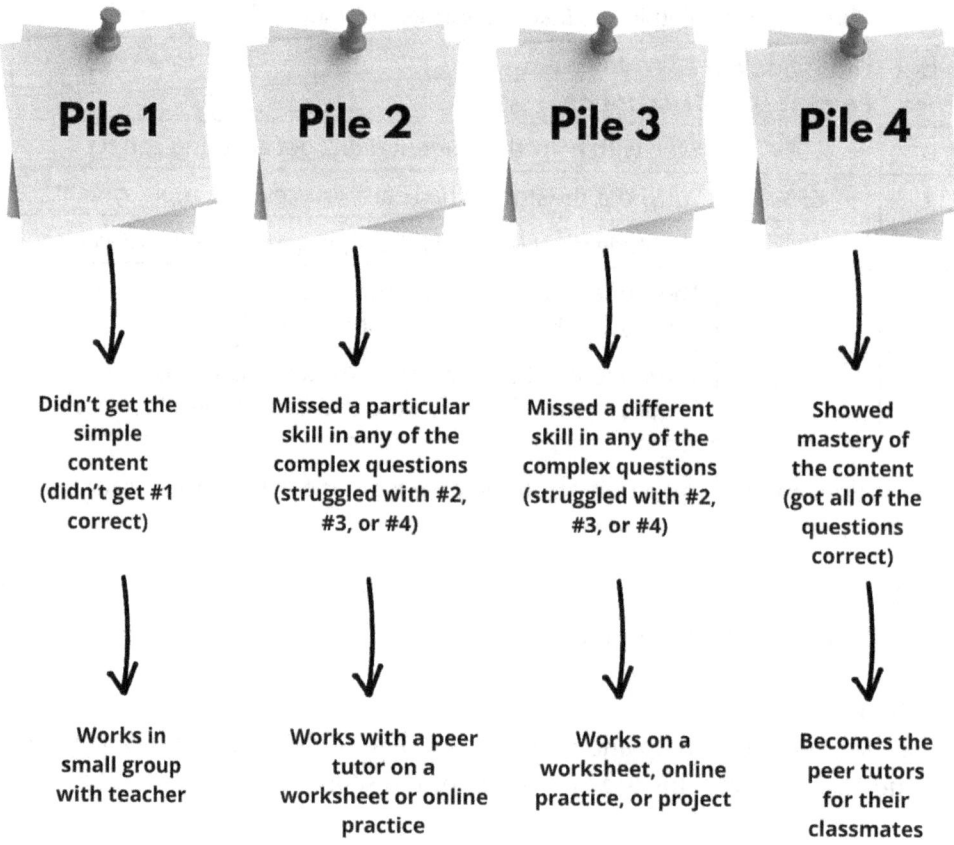

Figure 7.4 Sorting The Weekly FA Into Four Different Piles For Easy Small Group Assignments

the benefits of formative assessment and catch gaps in real time. While grading your weekly FA, separate into four piles and follow the flow in Figure 7.4 for each group.

Fast Feedback

The small groups should happen as soon as possible. For example if you give your weekly FA on Friday and pass back your FAs on Monday, the small groups also happen Monday. It's important that feedback is timely and that you implement the small groups as soon as possible so that students don't fall further behind as you add new content and skills.

How Often

Doing data driven small group instruction will benefit your students greatly, but it does take time to learn, finesse, and implement. If you are overwhelmed a good goal would be to do data driven small groups once every month. If you like it and see the benefits in your students, increase the frequency. I personally used data driven small groups once or twice a month on Monday's when I passed back my students weekly FA. In this section I'll take you through how to use the weekly FA to collect data that drives the small group formation and content selection. The actual time in small groups doesn't need to take forever, using twenty minutes on a Monday is plenty of time.

Student Feedback Tracking

After grading the weekly FA, it's essential to provide timely feedback to students in order to increase their ownership and buy in to the assessment process. Students need to be reminded each week of their progress toward mastery so that there are no surprises when they (or their parent) see their grade in the gradebook. You should pass back assessments the next day whenever possible to keep the connection between what they are learning and their mastery progression. Use a student tracker that enables students to shade in their rubric scores on a bar graph for an entire unit so they can track their mastery. Providing this visual track of their progress is helpful for students to quickly see how they are doing overall for the unit. You can download my student tracker in the resources section of www.gatebreakerbook.com.

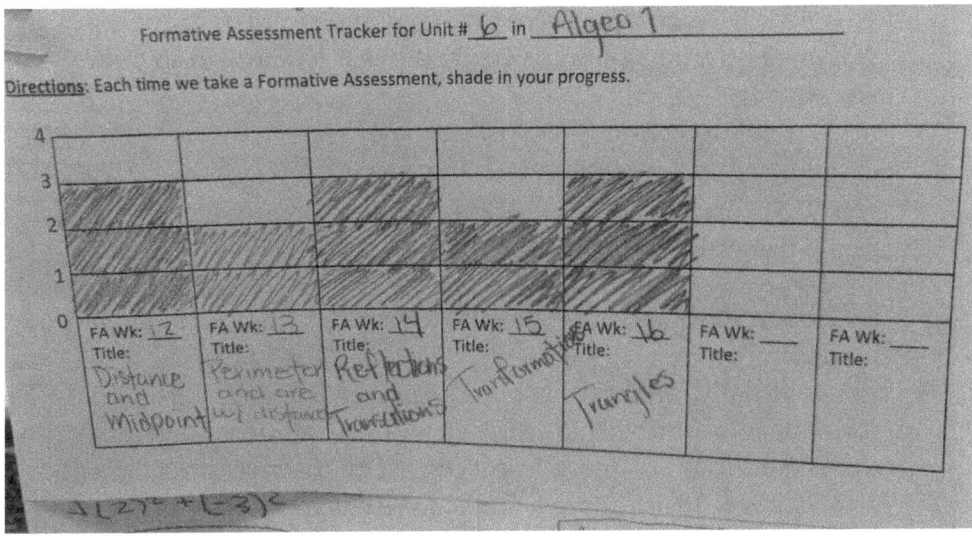

Figure 7.5 Weekly FA Student Tracker
Photography by author.

Part 3: Summative Assessment

The biggest question I get about this assessment framework is, "can I use this grading style if I'm the only teacher grading this way and I need to use common assessments with my department or district?" The answer is 100% yes, you can take any curriculum created assessment, department collaborative assessment, or district benchmark and apply these strategies to grade it equitably using a mastery-based approach and I'm going to show you how in this final part of the Rethinking Math Assessment Framework™. In Part 3 we'll talk about aligning your summative assessment, how to grade the summative assessment, and student reflection and revision with the summative assessment.

Summative Assessment Alignment

You do not need to alter any existing summative assessment to grade it using a mastery-based approach. You can use any chapter test, unit test, department common assessment, or district exam and still grade with this system. Here is what you'll need to do to make sure your assessment is aligned and ready for success with the mastery-based grading system:

- ◆ *Step 1:* Categorize all questions as simple, complex, or challenge.
- ◆ *Step 2:* Count up how many simple, complex, and challenge questions your test has. Make sure there are more complex questions than simple questions (it'll make grading easier, trust me). If your summative assessment didn't have any challenge questions, find or create at least one to add to your assessment.
- ◆ *Step 3:* Total up your question categories.
  - How many simple (Level 2) questions do you have? Write it down.
  - How many complex (Level 3) questions do you have? Write it down.
  - How many challenge (Level 4) questions do you have? Write it down.
- ◆ *Step 4:* Visually identify questions as simple or complex on the assessment for students to see. This will help students understand which questions they need to get correct on the assessment to get them closer to their ultimate grade goal which creates more buy in and motivation on the assessment.

Summative Assessment Grading

You'll grade the assessment using the guidelines and methodology in the table from Part 2 for the weekly assessment as well as the same rubric to percent conversion table as you did for the formative assessments.

- *To score a 2*: Students will need to get the majority of the Level 2 questions correct. For example, if your test has eight simple questions, students will need to get five of those specific questions correct to get a score of 2 on the assessment
- *To score a 3*: Students will need to get almost all of the simple questions correct plus the majority of the Level 3 questions correct. For example, if your test has seven complex questions, students will need to get four of those specific questions correct on top of getting almost all of the simple questions correct in order to get a score of 3 on the assessment
- *To score a 4*: Students will need to get all of the simple questions correct, almost all of the complex questions correct, plus the majority of the Level 4 questions correct. For example, if your test has three challenge questions, students will need to get two of those specific questions correct on top of getting all of the simple questions correct and most of the complex questions correct to get a score of 4 on the assessment. It's very common for tests to have just one challenge question and that's okay. Students will just need to get that one question correct (on top of the simple and complex questions as well) to score a 4.

Student Reflection And Revision

It is important to help students learn from their mistakes and reflect on their performance. I encourage all teachers to take the day they pass back the summative assessment to include 15–30 minutes for students to reflect on their scores. Ideas of what to include on a summative assessment reflection include:

- Do you feel your grade is fair?
- What grade would you like to get on your next assessment?
- What will you need to adjust about your class time, study habits, etc. in order to achieve that grade on the next assessment?
- Space for test corrections
- Space to explain the correction they made on any incorrect problems

With mastery-based grading you must also allow students to retake summative assessment at any point. Mastery can happen at any time and should always be accurately reflected in the student's grade. Students' lower grade should always be fully replaced with a grade that correctly communicates their mastery of the given standards or content. You do not need to give class time for students to retake assessments; you've already given them class time and if they want to retake an assessment to show mastery they can make arrangements with you to come at lunch or after school.

Discussion

There's no doubt about it, this grading system is different from what many teachers are used to, but also it's different from what many students are used to. I recommend sharing the ins and outs of mastery-based grading with your students in order to reap all of the benefits of student motivation, perseverance, and buy in to this system. Helping students understand that their grade is no longer about how many questions they get correct, but instead about the type of questions they get correct (simple, complex, or challenge) will be a shift for them too. In order for this system to be a powerful motivator, students have to deeply understand how it works. Ideas for explaining this system to your students include:

◆ Encourage students to share their experiences of feeling unfairly graded in the past through pair-sharing or group discussions. This opens up a conversation about perceptions of fairness, transparency, and equity in grading.

◆ Share the weekly FA structure with them explaining that all weekly FAs will have one simple question, three complex questions, and one challenge question.

◆ Share the Rubric To Percent Rethinking Math Assessment Framework™ Conversion Table with them and get a discussion going about what they notice and what they wonder about the information on it.

◆ Remind students that with mastery-based grading they can always retake any assessment and the higher grade will replace the lower grade.

◆ Make sure you point out the 50% floor and discuss how struggling on one assessment will not make it impossible to raise your grade.

Helping students understand that their grade is no longer about how many questions they get correct, but instead about the type of questions they get correct (simple, complex, or challenge) will be a shift for them too.

Shifting your assessment practices to a mastery-based grading approach is a huge jump. I'm well aware of that. When I was required by administration to switch to grading with a rubric I did not want to do it and as I shared in the opening of this chapter, two teachers at my school even got fired because they refused to switch to this system. How we grade and assess students feels oddly personal for some reason and it's normal to experience dissonance and ambivalence about making the jump to mastery-based grading. However, you picked up this book for a reason. You picked up this book because you want to make a difference in the lives of your students who need you the most, who struggle with math the most, and who might be struggling with school the most. You picked up this book because you believe in your students – all students – and want them to experience success in mathematics. You can follow all of the other sections of this book, but if you leave this part out, your classroom will not make the full transformation you seek. Student motivation will come when, and only when, students feel they are being graded equitably in your classroom. Without this piece, students will still struggle to understand why all the work is worth it if they don't see movement in their grade. You can learn even more about this grading system and get additional support in the chapter resources section at www.gatebreakerbook.com. If you purchased multiple copies of this book for your team, you can access additional bulk order bonuses for the Rethinking Math Assessment Framework™ when you visit www.teams.gatebreakerbook.com.

If you're not sold on switching to mastery-based grading I encourage you at the very least to reflect on assessment and grading in your classroom. Journal about the following prompts and if equity isn't at the center of your responses, get curious about that. Research alternative grading systems that feel right to you and are equitable for students.

Assessment Reflection Questions

1 What are your assessments measuring? Student understanding? Repetition? Mimicking?
2 What feedback does your grade provide to students about their understanding of the math concepts?

3 Is it possible for students to recover from a poor score or two? What would they need to do to recover to a C or higher?

4 Are your assessment and grading practices motivational for students who struggle?

36% Increase In Proficiency

I was struggling with student engagement in my 7th grade math class. Prior to implementing Juliana's methods, particularly the assessment approach and the Math Wars Method®, I had only 56% of my students scoring proficient on our state test and I had zero students score at level five, the highest level of the test.

I registered for Juliana's online workshops because I was interested in grading based on students' proficiency of the standards, and not just the number correct out of the total. I wanted to see how using this type of grading would motivate students on their assessments and if it could do anything to help with engagement.

Since implementing the framework I have a more holistic grading approach. Students seem to be more motivated to achieve proficiency as opposed to just getting a grade, which I love. In my district, we use a 0–100% grading system so I had to use the chart that transfers the grade from a rubric score to a percentage. I give the five question quiz almost every Friday and it really tells me how well students mastered the material from the week. Plus, the Math Wars Method® is a game changer. It gives me so much information during a lesson and allows me to adjust or help individual students right at their seat. It is so much more effective than just teaching the notes.

I just received my students' state test results and 92% of my students scored proficient or above! In fact, 39% of my students scored a level five on the state test. My students are more motivated in class and on assessments than I could have imagined. The students are so much more engaged.

Benny, Middle School Math Teacher in Florida

✅ Gatebreaker Homework

◆ Implement the Rethinking Math Assessment Framework™ and begin grading students using a mastery-based approach

◆ If you're not ready to implement the Rethinking Math Assessment Framework™, reflect on the questions from the *Assessment Reflection Questions* section at the end of the chapter

You stuck with it and finished the entire B.R.E.A.K. it™ Math Intervention Framework. Once you implement each phase and each step of this framework I know you will see the student growth you were hoping for when you first opened this book. Let's spend just a little more time together in the last chapter of this book to come full circle on a few things.

Reference List

Marzano, R. (2009). *Formative Assessment and Standards-Based Grading*. Marzano Research.

William, D. (2011). *Embedded Formative Assessment*. Solution Tree.

8

Conclusion: Do Something Different

B.R.E.A.K. it™ Math Intervention Framework
helping teachers break the gatekeeping cycles of mathematics

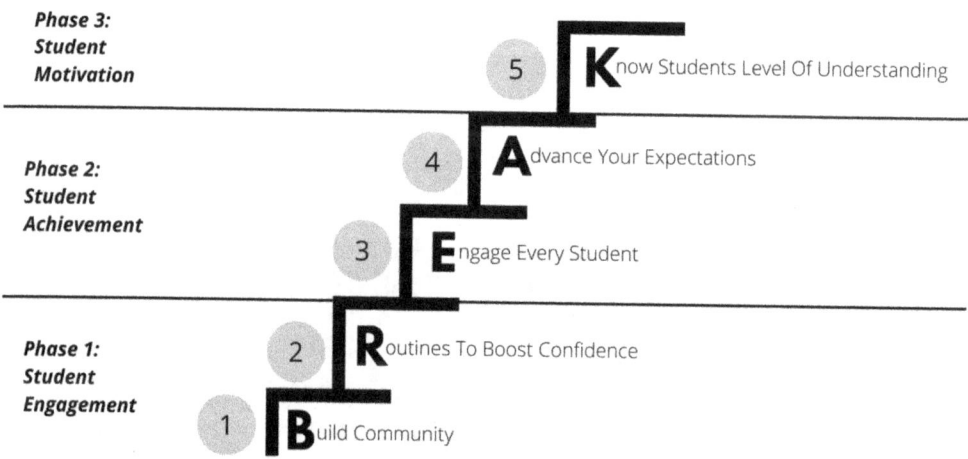

Figure 8.1 The B.R.E.A.K. it™ Math Intervention Framework

You've made it to the end and I'm so proud of you. To be honest I don't actually finish many books unless I have found immense value in them. So to think that you've found immense value in this book means the world to me. Another thing that would mean the world to me since you'd made it this

DOI: 10.4324/9781003479703-12

far is to take one minute to leave a review of this book on Amazon when you finish this chapter.

This book has taken you on a journey and an exploration of the B.R.E.A.K. it™ Math Intervention Framework as you begin to break the gatekeeping cycles of mathematics. We started in Phase 1, student engagement, where we learned the importance of building intentional community for students who struggle with math as well as the Six Core Engagement Structures that boost student confidence. In this phase students should be voicing their math anxiety and working on building a new relationship with mathematics in a safe classroom community in order to free up working memory for retention and recall. Next was Phase 2, student achievement, where we learned how to engage every student with the Math Wars Method® and the importance of utilizing a just in time math intervention approach to advance our expectations of students. In this phase students are growing in their confidence of being doers of mathematics and rising to meet your high grade level expectations. Lastly, we reached Phase 3, student motivation, where we learned the Rethinking Math Assessment Framework™ that allows us to know our students' level of understanding. In this phase students are becoming more intrinsically motivated to achieve at high rates and have more buy in to the connection between their effort and the result in the gradebook.

However, this book and these strategies will only work if you actually do something different as a result of reading them so we're going to spend our final pages together clearing any roadblocks that might be in your way. If you purchased multiple copies of this book for your team and are seeing additional copies for your school or district, please contact specialsales@taylorandfrancis.com for discount inquiries.

Connecting Back

For us to close out this book together I need you to go back to where we first began, your "so that" statement in Chapter 1. Your "so that" statement is what comes after you tell someone, "I teach math." *I teach math so that... what?* Go back to Chapter 1 and re-read what you wrote. Is your "so that" statement the same or has it changed as a result of what you've read? Write a new one if you need to that brings in your new awareness and motivation for teaching students who have been historically unsuccessful in math.

Next we need to identify and overcome the potential roadblocks in the way of us truly becoming gatebreakers. Go through the following reflection prompts inspired by the "Jobs To Be Done Theory" from work by Christensen (2016).

Your "so that" statement

Copy what you wrote in Chapter 1 or create a new "so that" statement based on what you've learned in this book.

Identifying Opposition: Are there any parts of the B.R.E.A.K. it™ Math Intervention Framework that seem in opposition to the things you love most about your classroom/teaching? If so, list them here.

Fears: What are your biggest fears about implementing the framework in your classroom? Be specific here. Which parts are you most hesitant or fearful to implement?

Take One Action: What is just one habit you can change to begin implementing the framework in your classroom? (If you feel lost or overwhelmed, might I suggest the very first gatebreaker homework assignment in Chapter 3)

Thank you for going on this journey with me. Thank you for wanting to engage your students who have been historically unsuccessful in math and finally help them thrive in mathematics. Mathematics has a huge impact on students and their futures and as a math teacher, you hold immense power in the lives of your students. You can help put them on the path to an open gate and achieving at high levels, or a closed gate and struggling through their educational and even professional career. I believe you are already an amazing teacher and that your students are lucky to have you. Now take these tools, strategies, and activities and become a gatebreaker!

⊘ One Final Gatebreaker Homework Assignment

- ◆ Reflect on the questions above
- ◆ Take one minute to write an honest review of this book on Amazon
- ◆ Connect with me in one of the ways identified below

If you want more support on your journey to becoming a gatebreaker, here are some more ways we can work together:

- ◆ **Digital PD offerings:** I'm constantly changing, updating, and improving how I support teachers and schools in this important work. I offer a variety of online professional development workshops, memberships, and free experiences that can be found at www.collaboratedwithjuliana.com/pd.
- ◆ **School and district professional development and coaching:** Each year I work in depth with a few schools and districts to help them increase math outcomes for 6–12th grade students who struggle. Send me an email juliana@collaboratedconsulting.org to schedule an initial meeting to see if we are a good fit or visit my website www.collaboratedconsulting.org/consulting.
- ◆ **Join my weekly email list:** Each Wednesday I send tips, resources, discounts, and free workshops to teachers and administrators within my email community. Visit www.collaboratedwithjuliana.com/subscribe to join the fun!

Reference List

Christensen, C. M., Hall, T., Dillon, K., & Duncan, D. S. (2016). *Competing against luck: The story of innovation and customer choice.* Harper Collins.